CAMBRIDGE STUDIES IN
ADVANCED MATHEMATICS II

Local representation theory

Local representation theory

Modular representations as an
introduction to the local representation
theory of finite groups

J. L. ALPERIN

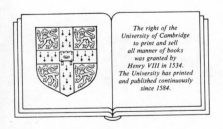

The right of the
University of Cambridge
to print and sell
all manner of books
was granted by
Henry VIII in 1534.
The University has printed
and published continuously
since 1584.

CAMBRIDGE UNIVERSITY PRESS

Cambridge
London   New York   New Rochelle
Melbourne   Sydney

Published by the Press Syndicate of the University of Cambridge
The Pitt Building, Trumpington Street, Cambridge CB2 1RP
32 East 57th Street, New York, NY 10022, USA
10 Stamford Road, Oakleigh, Melbourne 3166, Australia

© Cambridge University Press 1986

First published 1986

Printed in Great Britain at the University Press, Cambridge

British Library cataloguing in publication data
Alperin, J. L.
Local representation theory: modular
representations as an introduction to the local
representation theory of finite groups.–
(Cambridge studies in advanced mathematics; 11)
1. Modular representations of groups. 2. Finite
groups
I. Title
512′.2   QA171

Library of Congress cataloging-in-publication data
Alperin, J. L.
Local representation theory: modular
representations as an introduction to the local
representation theory of finite groups.
(Cambridge studies in advanced mathematics; 11)
Includes index.
1. Finite groups. 2. Modular representations of
groups. I. Title. II. Series.
QA171.A545   1986   512′.2   85-17436

ISBN 0 521 30660 4

TO
GREGORY AND COURTNEY

Contents

Preface

The representation theory of finite groups is at the same time an old and well-developed subject and a quickly changing and dynamic one. This may be daunting to many readers but we propose a remedy: in this short text, despite the great breadth of the topic, we shall take the reader to one of the high points of the subject. Hence, we shall go very far very quickly. This emphasis on speed and height suggests an analogy with climbing a mountain and doing it in quite short order. Carrying this analogy further, it means that we must not burden ourselves down with lots of heavy equipment. This means, on the one hand, that we can move quickly up the mountain but that, on the other hand, there will be times where we will not have the best tools available. Specifically, in our case, we are going to limit ourselves to a single set of methods, even though there are several, no one of which is sufficient to achieve all the results, even the most important ones.

Local representation theory is the part of the subject that involves and relates representations in characteristic zero and in characteristic p, p a prime. This is analogous with many ideas in number theory. Local representation theory is also local in a second sense: the p-local subgroups play a central role. These subgroups, the normalizers of the non-identity p-subgroups, are critical here (as they are all over group theory) and arise in all the basis theorems.

Our plan is to study only kG-modules, where G is a finite group and k is an algebraically closed field of characteristic p. We shall not go into the characteristic zero consequences or the applications to group structure. In this way we can go very deeply in little time. After two introductory chapters, we shall prove the basic results of Green and use them, following along the lines of our work with Burry, to prove the fundamental results. In particular, we shall use the theorem of Burry–Carlson–Puig to establish

Brauer's First Main Theorem and then prove the module form of Brauer's Second Main Theorem due to Nagao. We shall then treat Feit's results connecting maps and the Green correspondence with a new module-theoretic argument. We shall conclude with the Brauer–Dade cyclic theory proceeding mainly along the lines of the Green–Peacock approach, except that the last few sections are new.

This is a text, so the reader will not find the completeness or best results that he would expect in a treatise, nor will he find any historical account. As a text this book is suitable for a one-semester course, with some deletions, or a year course with the addition of material on the relations between representation theory in characteristic p and zero and on the applications. The exercises have been carefully selected and a few are used later in the text.

We hope we can launch the reader into this broad and exciting subject. All of representation theory is divided into three parts. First, there is the general theory, second the representation theory of the most important groups, like the Lie type groups, and third there are connections with many other areas in that these other fields are used in studying representations and there are applications to these areas. These include structure of groups, number theory, algebras, homological algebra, combinatorics, orders, algebraic geometry and algebraic groups. We hope we have given an introduction which will stimulate the reader to explore all these wonderful ideas.

I

Semisimple modules

Simple modules and simple algebras give a great deal of information about arbitrary algebras and their modules and it is this point that we shall develop in this chapter. Semisimple modules are the key idea. They provide a quick proof of the celebrated Wedderburn theorem on simple algebras and they show how any module can be viewed as consisting of many layers of semisimple modules. Our main interest is group algebras and we shall see quickly how these general results lead to insights about group representations.

Let us establish some fixed notation. We let A be a finite-dimensional algebra with unit element over an algebraically closed field k whose characteristic is p. All A-modules are assumed to be left modules and finite-dimensional over k. We also fix a finite group G and let kG be the group algebra of G over k. When we refer to p-groups or p-subgroups of G we shall be implicitly assuming that p is a prime.

1 Simple modules

There is a close connection between the structure of A and the structure of A-modules. Certainly the most important A-module is A itself with the module structure given by left multiplication. Moreover, if U is any A-module and $u_1, \ldots, u_n$ are generators for U (for example, a basis of U over k) then U is a homomorphic image of the A-module $A \oplus \cdots \oplus A$, the direct sum of n copies of A, via the map which sends the n-tuple $(a_1, \ldots, a_n)$ to $a_1 u_1 + \cdots + a_n u_n$, as is easily verified. If U is a simple A-module then any non-zero element of U is a generator so that U is a homomorphic image of the A-module A. This implies that there are only finitely many simple A-modules, up to isomorphism, as the A-module A has a composition series and we have just seen that any simple module is a composition factor.

We shall examine modules built up from simple modules by using direct sums. This will be the basis for all our important results.

Lemma 1 *If the A-module U is a direct sum of the simple submodules $S_1, \ldots,$ S_n and V is a submodule of U then there is a subset I of $\{1, \ldots, n\}$ such that U is the direct sum of V and the sum of all the S_i, $i \in I$.*

We shall use the usual notation S_I for the latter sum, where $S_\varnothing = 0$. In order to prove the result, simply choose a subset I maximal subject to the condition that S_I and V intersect in 0. We need only see that U is the sum of V and S_I. But, if this is not the case then there is j, $1 \leqslant j \leqslant n$ with S_j not contained in $S_I + V$. Thus, as S_j is simple, we have $S_j \cap (S_I + V) = 0$ and $S_j + S_I + V$ is also a direct sum, contradicting the maximality of I.

Proposition 2 *If U is an A-module then the following are equivalent:*
  (1) *U is a direct sum of simple A-modules;*
  (2) *Every submodule of U is a direct summand.*

We shall call such modules *semisimple*. Our results imply some natural properties of this class of modules. First, the lemma gives us that a submodule of a semisimple module is again semisimple. Indeed, with the notation of the lemma, V is isomorphic with U/S_I which is, in turn by our assumption, isomorphic with S_J where J is the complement to I in $\{1, \ldots, n\}$. Second, any quotient of a semisimple is also semisimple. In fact, again with the same notation, U/V is isomorphic with S_I. This fact is also a consequence of the proposition. Indeed, let us see that the quotient U/V has the second property of the proposition. Let W/V be a submodule of U/V, where W is a submodule of U containing V. By the lemma, U is the direct sum of W and a submodule X. It follows that U/V is the direct sum of W/V and $X + V/V$: their sum if U/V and $W \cap (X + V) = V$ as $V \subseteq W$ and $W \cap X = 0$. Finally, it is trivial that the direct sum of semisimple modules is again semisimple.

It remains now to prove the proposition. In view of the lemma, it suffices to demonstrate that the second assertion implies the first. However, this second property passes to quotient modules: if W/V is a submodule of U/V, where $W \supseteq V$ are submodules of U, then U is the direct sum of W and X so U/V is the direct sum of W/V and $X + V/V$. (Indeed, $W + (X + V) \supseteq W + X = U$ and $W \cap (X + V) = V$ as $W \supseteq V$ and $W \cap X = 0$). This allows us to proceed by induction on the composition length of U. In fact, if S is a simple submodule of U then there is a direct sum $U = S + T$ so $T \cong U/S$ is a direct sum of simple modules, by induction, so certainly U has this property.

We are now going to apply these ideas to the algebra A. This is done by means of the *radical* of A, written rad A, which consists of the elements of A which annihilate every simple A-module, that is annihilate each semisimple A-module. (Recall that an element a of A annihilates the module M if $aM = 0$.) The radical of A is an ideal: it is closed under addition and if $a, b \in A$, $r \in$ rad A and S is a simple module, then

$$(arb)S = ar(bS) \subseteq arS = 0.$$

There are some remarkable and important characterizations of the radical.

Theorem 3 *The radical of A is equal to each of the following:*
  (1) *the smallest submodule of A whose corresponding quotient is semisimple;*
  (2) *the intersection of all the maximal submodules of A;*
  (3) *the largest nilpotent ideal of A.*

These statements require some explanation. The submodules of the A-module A are, of course, just the left ideals of A. It is part of the theorem that the submodule described in the first part actually exists. An ideal N of A is nilpotent if there is a positive integer n such that $x_1 x_2 \cdots x_n = 0$ whenever $x_1, \ldots, x_n \in N$. We are claiming also that there is a largest such ideal.

Before proving the theorem, let us see how it enables us to define a vitally important class of algebras. We say that the algebra A is *semisimple* if rad $A = 0$. The first part of the theorem shows that A is semisimple if, and only if, the A-module A is semisimple. Since every A-module is a homomorphic image of a direct sum of copies of A, it is immediate that A is semisimple exactly when all A-modules are semisimple. One reason this type of algebra is useful is that if A is any algebra then $A/$rad A is a semisimple algebra. Indeed, $A/$rad A is a semisimple A-module, by the theorem, so it is certainly still semisimple as an $A/$rad A-module. In fact, an A-module is semisimple if, and only if, it is an $A/$rad A-module regarded as an A-module, since each semisimple A-module is annihilated by rad A while all $A/$rad A-modules are semisimple. In particular, $A/$rad A has zero radical.

Let us now turn to the proof of this important result. Suppose that I and J are nilpotent ideals of A so that $I^m = J^n = 0$ for suitable positive integers m and n. Certainly $I + J$ is also an ideal and it is nilpotent: $(I + J)^{m+n} = 0$.

Indeed, if $x_i \in I$, $y_i \in J$, $1 \leqslant i \leqslant m+n$, then

$$(x_1 + y_1) \cdots (x_{m+n} + y_{m+n})$$

is a sum of 2^{m+n} terms each of which has at least m factors from I or at least n factors from J. However, I and J are ideals, so each term is a product of m elements from I or n elements from J. Thus, $(I+J)^{m+n} = 0$, the sum of nilpotent ideals is again the same, and there is a largest nilpotent ideal N of A.

If S is a simple A-module then NS is a submodule of S. Hence NS is S or 0. If $NS = S$ then $N^n S = S$ for any positive integer n, contradicting the nilpotence of N, so we deduce that $NS = 0$ and so $N \subseteq \text{rad } A$. To see that rad $A = N$ it now suffices to show that rad A is nilpotent as then rad $A \subseteq N$. Let

$$A = A_0 \supset A_1 \supset \cdots \supset A_r = 0$$

be a composition series for the A-module so each quotient A_i/A_{i+1}, $0 \leqslant i < r$, is simple. Hence, (rad $A)A_i \subseteq A_{i+1}$ so (rad $A)^r A_0 = 0$ and (rad $A)^r = 0$, as required.

Suppose that $M_1, \ldots, M_r$ are maximal submodules of the A-module so each quotient A/M_i is a simple module and $A/M_1 \oplus \cdots \oplus M_r$ is semisimple. It follows that $A/M_1 \cap \cdots \cap M_r$ is also semisimple since it is isomorphic with a submodule of the preceding direct sum (via the map sending the coset containing $a \in A$ to the r-tuple $(M_1 + a, \ldots, M_r + a)$). Hence, if I is the intersection of all the maximal submodules of A then A/I is semisimple because I is the intersection of a finite number of maximal submodules. On the other hand, suppose M is a submodule of A with A/M semisimple, in fact, say $A/M = L_1/M + \cdots + L_s/M$ is a direct sum when each L_i is a submodule containing M with L_i/M simple. Thus, if we let M_i be the sum of all the L_j, $j \neq i$, then $A/M_i \cong L_i/M$ is simple so M_i is maximal and we also have $M = M_1 \cap \cdots \cap M_s$. Thus, $M \supseteq I$ and so I is the smallest submodule of A with semisimple quotient. We have established that the submodule described in (1) exists and coincides with the submodule I. Hence, it remains only to prove that $I = \text{rad } A$.

However, if $a \in \text{rad } A$ and M is a maximal submodule then a annihilates the simple module A/M so $aA \subseteq M$ and, in particular, $a = a \cdot 1 \in M$. Hence, rad $A \subseteq M$ for each such submodule M and rad $A \subseteq I$, their intersection. Suppose that this inclusion is proper so I is not contained in rad A. Hence, there must be a simple A-module S with $IS \neq 0$. Choose $0 \neq s \in S$ with $Is \neq 0$ so $Is = S$ as Is is a submodule of S. Hence, there is $x \in I$ with $xs = -s$ so $(x+1)s = 0$. Therefore, $x + 1$ lies in a proper left ideal of A, the annihilator of s, so $x + 1$ lies in a maximal submodule M which contains this annihilator.

But $x \in I \subseteq M$, by definition of I, so $1 = (x+1) - x \in M$, a contradiction. Hence, rad A is not properly contained in I and the theorem is completely proved.

We shall now pause and examine a good example, namely $T_n(k)$, the algebra of $n \times n$ lower triangular matrices. Let N be the subset consisting of matrices with all diagonal entries zero. It is easy to verify that N is an ideal and that it is nilpotent; in fact, $N^n = 0$. Let M_i, $1 \leq i \leq n$, consist of the elements of $T_n(k)$ which have a zero in the ith diagonal position. It is also easy to see that M_i is an ideal. In particular, M_i is a submodule and, since it is of codimension one, it follows that $S_i = T_n(k)/M_i$ is a simple module. The way that an element t of $T_n(k)$ acts on S_i is by multiplication by that scalar which is the ith diagonal entry of t: this is immediate from matrix multiplication. Hence, $S_1, \ldots, S_n$ are n non-isomorphic simple $T_n(k)$-modules. We also have that $N = M_1 \cap \cdots \cap M_n$ so $T_n(k)/N \cong S_1 \oplus \cdots \oplus S_n$ since there is an embedding of the left-hand side in the right and both sides are n-dimensional. Thus, $T_n(k)/N$ is a semisimple module so $N \supseteq \mathrm{rad}\, T_n(k)$. But N is nilpotent so $N \supseteq \mathrm{rad}\, T_n(k)$ and therefore $N = \mathrm{rad}\, T_n(k)$. Finally, if S is any simple $T_n(k)$-module then S is isomorphic with one of $S_1, \ldots, S_n$. Indeed, S is isomorphic with a quotient of the $T_n(k)$-module $T_n(k)$ so S is a composition factor of $T_n(k)/\mathrm{rad}\, T_n(k)$.

The remainder of this first section is devoted to using our results to study modules which are not semisimple by showing how they are made up of 'layers' of semisimple modules.

Proposition 4 *If U is an A-module then the following are equal:*
    (1) $(\mathrm{rad}\, A)U$;
    (2) *the smallest submodule of U with semisimple quotient;*
    (3) *the intersection of all maximal submodules of U.*

Part of the proof of Theorem 3 showed that, for the case $U = A$, the submodule described in the second statement exists and is equal to the intersection described in (3). The argument works equally well for an arbitrary A-module. It suffices therefore to prove that the modules given in (1) and (2) are the same.

However, rad A annihilates $U/(\mathrm{rad}\, A)U$ so the quotient is a module for $A/\mathrm{rad}\, A$. But $A/\mathrm{rad}\, A$ is semisimple so each of its modules is semisimple, as we have seen above, and so $U/(\mathrm{rad}\, A)U$ is certainly semisimple as an $A/\mathrm{rad}\, A$-module and so an A-module. Thus, to prove the proposition, we need only show that if U/V is semisimple then $(\mathrm{rad}\, A)U \subseteq V$. However, the semisimplicity of U/V means that rad A annihilates U/V, which is just what we want.

The submodule of U given by the proposition is called the *radical* of U and is denoted rad(U). Of course, if $U = A$ then this definition coincides with the prior one. Since rad(U) is also an A-module, it too lies in a radical and

$$\text{rad(rad}(U)) = (\text{rad } A)((\text{rad } A)U) = (\text{rad } A)^2 U.$$

We denote this by $\text{rad}^2(U)$ and define $\text{rad}^n(U)$ recursively so $\text{rad}^n(U) = (\text{rad } A)^n U$ and is the radical of $\text{rad}^{n-1}(U)$. Since a power of rad A is zero we have that $\text{rad}^r(U) = 0$ with $\text{rad}^{r-1}(U) \neq 0$ (letting $\text{rad}^0(U) = U$ for convenience) for some positive integer r which is called the *radical length* of U. The sequence of modules

$$U = \text{rad}^0(U) \supseteq \text{rad}^1(U) \supseteq \text{rad}^2(U) \supseteq \cdots$$

is called the *radical series* of U. Each of the successive quotients is semisimple and we have a description of U in terms of semisimple modules.

This description of U starts with U and proceeds with a sequence of smaller and smaller modules. We now shall give another similar description that works the other way around, from small submodules to large. The first result is analogous to Proposition 4.

Proposition 5 *If U is an A-module then the following are equal:*
   (1) *the set of u in U with (rad A)$u = 0$;*
   (2) *the largest semisimple submodule of U;*
   (3) *the sum of all the simple submodules of U.*

It is trivial that the set given by (1) is a submodule and, since it is annihilated by rad A, it is certainly semisimple. If V_1 and V_2 are semisimple submodules then so is $V_1 + V_2$, inasmuch as this sum is a homomorphic image of $V_1 \oplus V_2$, so the module described in (2) does exist and thus contains the module from (1). But rad A annihilates any semisimple submodule so the first two modules coincide. Finally, the sum of simple submodules is semisimple, by the above argument, and any semisimple module is the direct sum of simple modules so the result is proven.

The submodule just described is called the *socle* of U and is written soc(U). Now $U/\text{soc}(U)$ is a module so it, too, has a socle and we let $\text{soc}^2(U)$ be the submodule of U containing soc(U) since that $\text{soc}^2(U)/\text{soc}(U)$ is the socle of $U/\text{soc}(U)$. Hence, $(\text{rad } A)^2 \text{soc}^2(U) = 0$, since $(\text{rad } A) \text{soc}^2(U) \subseteq \text{soc}(U)$ and $(\text{rad } A) \text{soc}(U) = 0$. In fact, this characterizes $\text{soc}^2 U$, since $(\text{rad } A)^2 u = 0$ means rad A annihilates (rad A)u so (rad A)$u \subseteq \text{soc}(U)$ and thus rad A annihilates the coset of soc U containing u. This means that this coset is in $\text{soc}^2(U)$. Similarly, we can carry on and define, inductively, $\text{soc}^n(U)$ (letting $\text{soc}^0 U = 0$ for convenience) and this will consist of the elements of U annihilated by $(\text{rad } A)^n$. We have a *socle series*

$$0 = \text{soc}^0(U) \subseteq \text{soc}^1(U) \subseteq \text{soc}^2(U) \cdots$$

and the first positive integer r with $\text{soc}^r(U) = U$ is called the *socle length* of U. Again we have a description of U in terms of semisimple layers.

Exercises

1 Determine the radical and socle series of the $T_n(k)$-module $T_n(k)$.
2 Prove that
$$\text{rad}^i(T_n(k))/\text{rad}^{i+1}(T_n(k)) \cong S_{i+1} \oplus \cdots \oplus S_n,$$
$0 \leqslant i < n$, and that
$$\text{soc}^{i+1}(T_n(k))/\text{soc}^i(T_n(k)) \cong S_{n-i} \oplus \cdots \oplus S_{n-i}$$
where $0 \leqslant i < n$ and S_{n-i} appears $n-i$ times.
3 If U is an A-module then the radical length and socle length of U coincide (and this common length is the *Loewy length*).
4 If U is an A-module of Loewy length s then, for $0 \leqslant i \leqslant s$, $\text{rad}^i(U) \subseteq \text{soc}^{s-i}(U)$.

2 Simple algebras

We shall use our knowledge of modules to prove the basic structure theorems for simple and semisimple algebras. The idea is to pass from information on module structure to knowledge about endomorphisms and then algebras.

For any A-module U we let $\text{End}(U)$ be the algebra of all endomorphisms of U. We also let A° denote the opposite algebra to A, that is, the algebra with the same underlying set and linear structure as A but with new multiplication, $a \circ b = ba$.

Lemma 1 $\text{End}(A) \cong A^\circ$.

This is a trivial but vital fact about the A-module A. It is easy to prove. For each $a \in A$, let ρ_a be the linear transformation of A given by $\rho_a(x) = xa$, for $x \in A$. It is easy to verify that $\rho_a \in \text{End}(A)$ and $\rho_a \circ \rho_b = \rho_{ba}$ whenever $a, b \in A$. Moreover, if $\rho \in \text{End}(A)$ let $c = \rho(1)$ so, for any $x \in A$,

$$\rho(x) = \rho(x \cdot 1) = x\rho(1) = xc = \rho_c(x)$$

and $\rho = \rho_c$. Hence, the map which sends $a \in A^{\circ}$ to ρ_a is an algebra homomorphism of A° onto End(A) and it is one-to-one since $\rho_a = \rho_b$ implies that $a = \rho_a(1) = \rho_b(1) = b$.

The other key fact is Schur's lemma, which says that only the scalar multiplications are endomorphisms of a simple module:

Lemma 2 *If S is a simple A-module then* End(S) $= kI$.

Proof Let $\rho \in$ End(S) so ρ is certainly a linear transformation of S. Let $\lambda \in k$ be an eigenvalue of ρ so $\rho - \lambda I$ is a singular linear transformation and also an endomorphism. Thus $(\rho - \lambda I)S$ is a submodule of S and properly contained in S. Hence, this image is zero, $\rho - \lambda I = 0$ and $\rho = \lambda I$ as asserted.

Now that we understand endomorphisms of simple modules, let us go on to direct sums of modules. Suppose that the A-module U is a direct sum $U = U_1 + \cdots + U_r$. If $\rho \in$ End(U) and $u_j \in U_j$ then $\rho(u_j)$ has an expression given by the direct decomposition: let ρ_{ij} be the function from U_j to U_i which attaches the ith component of $\rho(u_j)$ to u_j. Now ρ_{ij} is then the composition of the natural injection of U_j into U, ρ and the projection of U onto U_i so $\rho_{ij} \in$ Hom$_A(U_j, U_i)$. Moreover, $\rho(u_j) = \sum_i \rho_{ij}(u_j)$. Hence, if $\rho(u_1 + \cdots + u_r) = v_1 + \cdots + v_r$, with the obvious notation, then we can express this in terms of matrix multiplication as follows:

$$\begin{pmatrix} \rho_{11} & \cdots & \rho_{1r} \\ \vdots & & \vdots \\ \rho_{r1} & \cdots & \rho_{rr} \end{pmatrix} \begin{pmatrix} u_1 \\ \vdots \\ u_r \end{pmatrix} = \begin{pmatrix} v_1 \\ \vdots \\ v_r \end{pmatrix}.$$

Let E consist of all $r \times r$ matrices whose i, j entry comes from Hom$_A(U_j, U_i)$ so it is clear that such matrices add and multiply to form an algebra. To each $\rho \in$ End(U) we have attached $M(\rho) \in E$. Furthermore, it is easy to check that every element of E so arises and the mapping M is an isomorphism. Summarizing, we have

Lemma 3 *If the A-module U is a direct sum*

$$U = U_1 + \cdots + U_r$$

of submodules then End(U) *is isomorphic with the algebra of all $r \times r$ matrices whose i, j entries come from* Hom$_A(U_j, U_i)$.

This is just a generalization of the usual way of attaching matrices to linear transformations: we have gone from k-modules, that is, vector spaces, to A-modules. We can now use these ideas to derive easily a deep consequence.

Theorem 4 *If A is simple then A is a matrix algebra.*

More specifically, there is a positive integer n and an isomorphism between A and the algebra $M_n(k)$ of all $n \times n$ matrices over k.

Proof Let S be a simple submodule of A and let U be the sum of all submodules isomorphic with S. In particular, U is semisimple and, furthermore, if we express U as a direct sum of n simple submodules, then each of these summands is isomorphic with S since U, and hence each homomorphic image of U, is a sum of modules isomorphic with S. Hence, Lemmas 2 and 3 imply that $\operatorname{End}(U) \cong M_n(k)$.

However, if $\rho \in \operatorname{End}(A)$ then $\rho(U) \subseteq U$ because if T is a submodule of A isomorphic with S then the homomorphic image $\rho(T)$ of T is either zero or also isomorphic with S. From the proof of Lemma 1, we know this means that U is a right ideal so it must now be an ideal. But A is simple so $A = U$ and

$$A^\circ \cong \operatorname{End}(A) = \operatorname{End}(U) \cong M_n(k).$$

Thus, any isomorphism of A° onto $M_n(k)$, composed with the taking of transposes, is the needed isomorphism.

We have also proved that A is semisimple in the midst of this proof. It is conversely true that $M_n(k)$ is simple and semisimple. Indeed, $M_n(k)$ has a simple module S consisting of column vectors of length n over k; the simplicity of S is immediate from the fact that $M_n(k)s = S$ for any $s \in S, s \neq 0$. Moreover, $M_n(k)$, as a module over itself, has a direct decomposition

$$M_n(k) = C_1 + \cdots + C_n$$

where C_i consists of the matrices which have zero entries outside the ith column. Each C_i is isomorphic with S so $M_n(k)$ is semisimple. Moreover, $M_n(k)$ has just one simple module, as any simple module would be a composition factor of the module $M_n(k)$. Now suppose that $0 \neq I$ is an ideal of $M_n(k)$. Let L be a simple submodule of I so $L \cong S$ and L is a module summand of $M_n(k)$. In particular, there is a module endomorphism φ_i of $M_n(k)$ such that $\varphi_i(L) = C_i$. By Lemma 1, φ_i is given by a right multiplication, $L \subseteq I$, an ideal, so $C_i = \varphi_i(L) \subseteq I$. Hence $I = M_n(k)$ as desired.

In passing, note that the integer n of the matrix algebra $M_n(k)$ is characterized as the number of summands when $M_n(k)$ is expressed as a direct sum of simple submodules, all of which are then isomorphic. This fact is useful when we are working with non-semisimple modules later.

Our next goal is to generalize these results to semisimple algebras, again using module decompositions and endomorphisms. To do this we need the concept of a direct sum of algebras, which is really two ideas, internal and external direct sums, in analogy with the similar ideas for modules.

If $A_1, A_2, \ldots, A_t$ are algebras, form

$$A_1 \oplus \cdots \oplus A_t$$

and make this into an algebra, by defining all operations in a component-wise fashion, the direct sum of $A_1, \ldots, A_t$. Let B_i consist of the t-tuples which are zero except perhaps in the ith component, so B_i is an algebra, $B_i \cong A_i$ and the direct sum is the vector space direct sum of its ideals $B_1, \ldots, B_t$. On the other hand, if the algebra A has ideals $A_1, \ldots, A_t$ and A is the vector space direct sum $A = A_1 + \cdots + A_t$ then

$$A \cong A_1 \oplus \cdots \oplus A_t$$

by the obvious map, as is quite easy to verify.

This idea arises in several ways, one of which follows.

Lemma 5 *If the A-module U is a direct sum of submodules*

$$U = U_1 \oplus \cdots \oplus U_t$$

and $\mathrm{Hom}_A(U_i, U_j) = 0$ *when* $i \neq j$, *then*

$$\mathrm{End}(U) \cong \mathrm{End}(U_1) \oplus \cdots \oplus \mathrm{End}(U_t).$$

Proof This is an immediate consequence of Lemma 3, which implies that $\mathrm{End}(U)$ is isomorphic with the algebra of $t \times t$ matrices with zero i, j entries if $i \neq j$ and i, i entries from $\mathrm{Hom}_A(U_i, U_i)$.

Theorem 6 *If A is a semisimple algebra then A is the direct sum of matrix algebras.*

Proof Suppose that A has exactly t simple modules $S_1, \ldots, S_t$, up to isomorphism of course. The semisimplicity of A implies that we can express the A-module A as a direct sum

$$A = U_1 + \cdots + U_t$$

where U_i is a direct sum of simple modules each isomorphic with S_i. If $\varphi \in \mathrm{Hom}_A(U_i, U_j)$ then $\varphi(U_i)$ is a submodule of U_j and isomorphic with a quotient module of U_i so if $i \neq j$ we must have $\varphi = 0$. Hence, Lemma 5 gives us that

$$\mathrm{End}(A) \cong \mathrm{End}(U_1) \oplus \cdots \oplus \mathrm{End}(U_t).$$

However, $A^\circ \cong \mathrm{End}(A)$ and each algebra $\mathrm{End}(U_i)$ is a matrix algebra, just as in the argument of Theorem 4, so we have the desired isomorphism here, too, in the same way.

The converse to Theorem 6 is also true. Suppose that A is the direct sum of its subalgebras $A_1, \ldots, A_t$ and that each A_i is a matrix algebra. As we have seen above, the A_i-module A_i is the direct sum of isomorphic simple submodules. However, if $i \neq j$ then a product of an element of A_j and one from A_i is zero so any A_i-submodule of A_i is certainly an A-submodule. Thus, using the decomposition of A_i, for each i, we have that the A-module A is semisimple, that is, A is semisimple.

Moreover, this shows that the simple A_i-module S_i is an A-module, where each A_j, $j \neq i$, annihilates S_i and that every composition factor of the A-module A is isomorphic with one of $S_1, \ldots, S_t$. Since A_i does not annihilate S_i we have that these t simple modules are the distinct simple A-modules. It is useful to remember, too, that the multiplicity that S_i occurs in a decomposition of A into the direct sum of simple modules is equal to the size of the corresponding matrix algebra and also to the dimension of S_i.

We shall conclude this section with the following additional observation: with the same notation, the decomposition

$$A = A_1 + \cdots + A_t$$

is the unique decomposition of A into a direct sum of matrix algebras. This is absolutely unique, not just up to an isomorphism! More can be proved and we shall do this: the ideals of A are exactly the sums of A_j, j running over a subset of the integers from 1 to t. This will imply that the A_i are exactly the minimal ideals of A and the same would hold for any such decomposition.

Let I be an ideal of A. Let $\mathscr{I}$ consist of all integers i such that S_i is isomorphic with a composition factor of the A-module I. It suffices to prove that $A_i \subseteq I$. Indeed, it then follows that $A_{\mathscr{I}}$, the sum of all such A_i, is contained in I, that the A-module $A/A_{\mathscr{I}}$ has no composition factor isomorphic with any S_i, $i \in \mathscr{I}$, while I, and hence $I/A_{\mathscr{I}}$, have only these factors. Thus, $I/A_{\mathscr{I}} = 0$, that is, $I = A_{\mathscr{I}}$. Hence, to conclude, let $i \in \mathscr{I}$ so, as A is semisimple, there is a simple summand L of I isomorphic with S_i. Moreover, L is a summand of A so the images of L under endomorphisms of the A-module A contain all the simple summands in a decomposition of A_i. Since $L \subseteq I$ and I is an ideal, we have that these images lie in I so $A_i \subseteq I$ as asserted.

Exercises

1 Let A_0 be an algebra not necessarily having a unit element. Prove that there is an algebra A with unit element 1 containing A_0 as a subalgebra of codimension one.

2 (cont.) Show that every ideal of A_0 is an ideal of A.

3   (cont.) Prove that if A_0 has no non-zero nilpotent ideal then the same is
     true for A. Deduce that A_0 then is a direct sum of matrix algebras (and in
     particular has a unit element).
4   Demonstrate that any automorphism of the algebra $M_n(k)$ is inner by
     using the fact that $M_n(k)$ has a unique simple module.
5   Show that the number of isomorphism classes of simple modules for the
     algebra A equals the number of matrix summands of $A/\mathrm{rad}\ A$. See what
     this means for the case $A = T_n(k)$.

3 Group algebras

We now turn to the case where our real interest lies, namely the
group algebra kG. We shall study this algebra with regard to the material in
the two preceding sections so we want to know whether kG is semisimple,
what the simple kG-modules are and how many there are.

We begin with the *trivial kG-module* which is, in reality, very important
but certainly trivial to exhibit. Consider k as a one-dimensional vector
space and make it into a kG-module by defining $g \cdot \lambda = \lambda$ whenever $g \in G$,
$\lambda \in k$. This is visibly a simple kG-module and we shall make immediate use of
it in our first result.

Theorem 1 *The algebra kG is semisimple if, and only if, the characteristic p
of k does not divide the order $|G|$ of G.*

First, suppose that p does divide $|G|$ so p is a prime. If kG were still
semisimple then the trivial kG-module k, being of dimension one, would
appear once in a direct decomposition of kG into a sum of simple kG-
modules. In particular, in any composition series of kG there would be
exactly one composition factor isomorphic with k. Let us see that this is not
the case.

The homomorphism of G to the identity group extends to an algebra
homomorphism of kG to the group algebra of 1, which we may identify with
the algebra k. The kernel is the *augmentation ideal* $\Delta(G)$ of kG and so consists
of all elements $\sum \alpha_g g$ of kG with $\sum \alpha_g = 0$. In particular, $\Delta(G)$ is a submodule
of kG. Moreover, the module $kG/\Delta(G)$ is isomorphic with the trivial module
k: for $1 \in kG$, $1 \notin \Delta(G)$ and $g \cdot 1 = 1 + (g-1) \in 1 + \Delta(G)$ whenever $g \in G$. On the
other hand, consider the element σ of kG which consists of the sum of all
group elements. Since p divides $|G|$, it follows that $\sigma \in \Delta(G)$. But it is
immediate that $g\sigma = \sigma$, for any $g \in G$, so $k\sigma$ is a submodule of kG also

isomorphic with the trivial kG-module k. When the series of submodules

$$kG \supset \Delta(G) \supseteq k\sigma \supset 0$$

is refined to a composition series, we have demonstrated that kG is not semisimple.

Let us suppose that p does not divide $|G|$ so $|G|$ is an invertible element of k. Let V be a submodule of the kG-module U; it suffices to prove that V is a direct summand of U. Let π be a linear transformation of the vector space U which is a projection onto V: π is the identity on V, $\pi(U) = V$ so U is the vector space direct sum of V and the kernel of π. If π also happened to be an endomorphism of U, that is, commutes with the action of each $g \in G$, then the kernel of π would also be a kG-module and we would have the desired result. What we shall do is modify π by an averaging procedure to obtain a new map with all the requisite properties.

To do this we work with linear transformations of U so that when we write the element g from G it will be clear if we are really referring to the linear transformation g induces on U. Set

$$\pi' = \frac{1}{|G|} \sum_{g \in G} g\pi g^{-1}.$$

Certainly, π' is a linear transformation of U, $\pi'(U) \subseteq V$ since

$$\pi'(U) = \frac{1}{|G|} \sum_{g \in G} g\pi g^{-1} U \subseteq \frac{1}{|G|} \sum_{g \in G} g\pi U = \frac{1}{|G|} \sum_{g \in G} gV = V$$

and also π is the identity on V, since for $v \in V$ we have

$$\pi'(v) = \frac{1}{|G|} \sum_{g \in G} g\pi(g^{-1}v) = \frac{1}{|G|} \sum_{g \in G} g(g^{-1}v) = \frac{1}{|G|} |G|v = v$$

as each $g^{-1}v \in V$ and π is the identity on V. Hence, it suffices to show that π' is a kG-homomorphism. However, if $h \in G$ and $u \in U$ then

$$\pi'(hu) = \frac{1}{|G|} \sum_{g \in G} g\pi g^{-1} hu = \frac{1}{|G|} \sum_{g \in G} hh^{-1} g\pi g^{-1} hu$$

$$= h \cdot \frac{1}{|G|} \sum_{x \in G} x\pi x^{-1} u = h\pi'(u)$$

since $h^{-1}g$ runs over G as g does.

Thus, we know that, in general, $\operatorname{rad}(kG)$ is not zero and it is natural to enquire as to the structure of $kG/\operatorname{rad}(kG)$. This is a direct sum of matrix algebras, the number of which is the number of simple kG-modules, up to isomorphism. There is a nice determination of this number.

Theorem 2 *The number of simple kG-modules equals the number of conjugacy classes of G of order not divisible by the characteristic of k.*

For example, if $p = 2$ then the number is the number of conjugacy classes of elements of odd order while if $p = 0$ then it is the number of conjugacy classes. We shall postpone the proof of this important result to the end of the section. The reason for doing this is that the argument uses techniques of no further relevance to our treatment: to relate simple modules and conjugacy classes it is necessary to go into the structure of kG and to use ideas from Lie algebras. We shall now explore the consequences of this theorem for p-groups, cyclic groups and $SL(2, p)$, the group of 2×2 matrices of determinant one over the finite field with p elements.

Corollary 3 *If G is a p-group then the trivial kG-module is the only simple kG-module.*

Proof The identity element is the only one of order not divisible by p so kG has a unique simple module. Moreover, the trivial module is simple.

Next, let G be a cyclic group of order n, where $n = p^{a}r$ with p not dividing r and let g be a generator for G. The polynomial $x^{r} - 1$ is separable, by our assumption, so there are r distinct roots in k, the rth roots of unity. If λ is one of these roots we form a one-dimensional vector space over k and let g^{i} act on it by multiplication by λ^{i}. Since $\lambda^{r} = 1$ implies $\lambda^{n} = 1$ this is a module, certainly it is simple and a different choice of λ will give a non-isomorphic module as g would act differently. The group G is abelian so each conjugacy class consists of one element and, as G is cyclic, there are exactly r elements of order not divisible by p and these form a subgroup of order r. Hence, we have described all the simple modules.

It is not necessary though to refer to the theorem for a count in this case of cyclic groups. The argument above shows that any simple kG-module is one-dimensional so g must act by multiplication by an nth root of unity. However, we have $x^{n} - 1 = (x^{r} - 1)^{p^{a}}$, since k has characteristic p, so every nth root of unity is an rth root of unity.

For our final example we set $G = SL(2, p)$. All aspects of the structure of G are important to group theorists but we shall be content here just to state one important fact: G has exactly p conjugacy classes of elements of order not divisible by p. Hence, when we have constructed p distinct simple kG-modules we will then have them all.

We regard G as consisting of matrices with elements from the prime field of k. Hence, the vector space V_2 of columns of length two over k is a kG-module, inasmuch as we can multiply an element of G and a column. We set

$$X = \begin{pmatrix} 1 \\ 0 \end{pmatrix} \quad Y = \begin{pmatrix} 0 \\ 1 \end{pmatrix}$$

so if

$$g = \begin{pmatrix} a & b \\ c & d \end{pmatrix}$$

then $gX = aX + cY$, $gY = bX + dY$. If we now form the polynomial ring $k[X, Y]$ in X and Y then the action of each $g \in G$ extends to an automorphism of the polynomial algebra. Let V_n be the subspace of $k[X, Y]$ consisting of homogeneous polynomials in X and Y of degree $n-1$. In particular, V_n is a kG-module and V_2 is as before. A basis of V_n consists of the monomials $X^{n-1}, X^{n-2}Y, \ldots, Y^{n-1}$, so V_n has dimension n. When $n = 1$ we have that V_1 is the one-dimensional vector space spanned by the unit element of $k[X, Y]$ and V_1 is the trivial kG-module. We assert that $V_1, V_2, \ldots, V_p$ are simple kG-modules. (The argument is really an adaptation to characteristic p of ideas about weights in representations of Lie algebras.)

Hence, let $1 \leqslant n < p$ and we shall prove that V_{n+1} is simple. We choose two elements from G, namely

$$g = \begin{pmatrix} 1 & 0 \\ 1 & 1 \end{pmatrix}, \quad h = \begin{pmatrix} 1 & 1 \\ 0 & 1 \end{pmatrix}$$

and we shall consider V_{n+1} as a module for the groups $\langle g \rangle$ and $\langle h \rangle$ generated by these elements. We shall prove that X^n is a generator of the $k\langle g \rangle$-module V_{n+1} and that the subspace kY^n spanned by Y^n is the socle. The same argument, with X and Y interchanged and with h used in place of g, will go right through and give that Y^n is a generator of the $k\langle h \rangle$-module V_{n+1} and kX^n is the socle. This will establish our assertion. Indeed, suppose that W is a non-zero submodule of V_{n+1}. Hence, W is certainly a $k\langle g \rangle$-submodule so contains a simple $k\langle g \rangle$-submodule. But there is only one simple $k\langle g \rangle$-submodule in V_{n+1}, namely kY^n, so $Y^n \in W$. Therefore, the $k\langle h \rangle$-module generated by Y^n is also contained in W, so $W = V_{n+1}$, as required.

In order to prove our claim about the structure of V_{n+1} as a $k\langle g \rangle$-module we shall demonstrate a stronger result which can be proved by an induction. For $0 \leqslant i \leqslant n$, let W_{i+1} be the $i+1$-dimensional subspace of V_{n+1} which has basis $X^i Y^{n-i}, X^{i-1}Y^{n-i+1}, \ldots, XY^{n-1}, Y^n$ and set $W_0 = 0$ so

$$V_{n+1} = W_{n+1} \supset W_n \supset \cdots \supset W_1 \supset W_0 = 0$$

with each subspace in this series of codimension one in the preceding one.

For each i, $0 \leqslant i \leqslant n$, we shall prove the following assertions:

(1) W_i is a $k\langle g \rangle$-submodule;

(2) W_i/W_{i-1} is a trivial $k\langle g \rangle$-module;

(3) each element of $W_i - W_{i-1}$ generates W_i as a $k\langle g \rangle$-module.

We shall prove this by induction on i and then see how this establishes the desired result for V_{n+1}.

Now $gX = X + Y$ and $gY = Y$, so $gY^n = Y^n$ and our claim is valid for $i = 1$. Suppose it holds for i and let us turn to W_{i+1}. Using the binomial expansion, we have

$$gX^iY^{n-i} = (X+Y)^iY^{n-i} = X^iY^{n-i} + \binom{i}{2}X^{i-1}Y^{n-i+1} + u$$

where u is a linear combination of $X^{i-2}Y^{n-i+2}, \ldots, XY^{n-1}, Y^n$, and is therefore in W_{i-1}. Since W_{i+1} is spanned by X^iY^{n-i} and W_i and the latter is a $k\langle g \rangle$-module this proves that W_{i+1} is also. Moreover, since W_{i+1}/W_i is one-dimensional and this calculation shows that $gX^iY^{n-i} - X^iY^{n-i}$ is in W_i, we also have that W_{i+1}/W_i is a trivial $k\langle g \rangle$-module. Finally, since $i < p$ we have

$$\binom{i}{2} \neq 0$$

so $(g-1)X^iY^{n-i}$ is an element of $W_i - W_{i-1}$. Thus, if $v \in W_{i+1} - W_i$ then

$$v = \alpha X^iY^{n-i} + w$$

where $\alpha \neq 0$ and $w \in W_i$ and so

$$(g-1)v = \alpha(g-1)X^iY^{n-i} + (g-1)w.$$

But $(g-1)w \in W_{i-1}$, as W_i/W_{i-1} is a trivial $k\langle g \rangle$-module, so $(g-1)v \in W_i - W_{i-1}$. Hence, the $k\langle g \rangle$-module generated by v contains w_i, as it contains an element of $W_i - W_{i-1}$, and it contains v so it is just W_{i+1}, as required, and our three claims are valid.

In particular, the element X^n generates V_{n+1} as a $k\langle g \rangle$-module, since $X^n \in W_{n+1} - W_n$. It remains to examine the socle of V_{n+1} as a $k\langle g \rangle$-module. However, $g^p = 1$, as it is trivial to calculate, so $\langle g \rangle$ is a p-group. Thus, the trivial $k\langle g \rangle$-module is the unique simple $k\langle g \rangle$-module, so the subspace of V_{n+1} of vectors left fixed by g is the socle. But if $v \in V_{n+1} - W_1$ then the $k\langle g \rangle$-module generated by v is not one-dimensional, so $gv \neq v$. Hence $W_1 = kY^n$ is the socle as claimed. Hence, the simple kG-modules are as described.

We now drop our assumption that G is a special group and turn to our first contact with Clifford theory, the body of theory relating normal subgroups and representation theory. We shall prove the most basic result, namely, Clifford's theorem. We need one standard piece of notation: if H is a subgroup of G and V is a kG-module then V_H is the kH-module which is the restriction to kH.

Theorem 4 *If U is a semisimple kG-module and N is a normal subgroup of G then U_N is semisimple.*

Proof Let S be a simple kG-module; it suffices to show that S_N is semisimple. First we make an observation: if V is a kN-submodule of S_N then so is gV, for any $g \in G$. Indeed, if $y \in N$ then

$$y \cdot gV = g \cdot g^{-1} y g V = gV.$$

Now, let T be a simple submodule of S_N. If $g \in G$ then gT is also a submodule of S_N and it is also simple: if U were a non-zero proper kN-submodule of gT then $g^{-1}U$ would be the same for $g^{-1}gT = T$. Hence, the sum of all submodules gT, as g runs over G, is a semisimple kN-module and it is visibly a kG-module so it must be S and S is semisimple.

Clifford's theorem gives us another way to prove Corollary 3. Let G be a p-group and S a simple kG-module. We shall show that S is trivial, by induction on $|G|$. Let N be a normal subgroup of index p in G so S_N is semisimple, by Clifford's theorem, and thus each element of N leaves fixed each element of S. Hence, S is, in an obvious and natural way, a $k[G/N]$-module, so, by our analysis of the cyclic case, S is a trivial kG-module.

We now turn, at last, to the proof of Theorem 2. It is suggested to the reader to omit reading this argument, at a first reading, since, while it is very clever, it is intricate and presents no techniques we shall have any further use for.

However, for those who wish to read on in this section, for the sake of motivation let us first deal with the case that kG is semisimple, where the argument is easy but sets the stage for Brauer's lovely proof of Theorem 2. To do this, we let the center $Z(A)$ of any algebra A be the subalgebra of the elements which commute with all elements. If follows immediately that if A is a direct sum of subalgebras then $Z(A)$ is the direct sum of their centers. In particular, if $kG = A_1 + \cdots + A_r$ is a direct sum of matrix algebras then $Z(kG) = Z(A_1) + \cdots + Z(A_r)$. However, the scalar matrices form the center of the algebra of $n \times n$ matrices, for any n, so if e_i is the unit element of A_i then $Z(kG) = ke_1 + \cdots + ke_r$ and $e_1, \ldots, e_r$ is a basis for $Z(kG)$. But r is also the number of simple kG-modules so this coincides with the dimension of $Z(kG)$. Since the characteristic does not divide the order of G, we wish to prove that r is also the number of conjugacy classes of G. Let $C_1, \ldots, C_s$ be these classes and let $\gamma_1, \ldots, \gamma_s$ be their sums so $\gamma_1, \ldots, \gamma_s$ are linearly independent elements of kG. Moreover, if $g \in G$ then $g^{-1}\gamma_i g = \gamma_i$, since we just have a rearrangement of terms, so $\gamma_i \in Z(kG)$. Conversely, if $a = \sum \alpha_x x \in Z(kG)$ then $\alpha_{gxg^{-1}} = \alpha_x$ for all x, g since $\alpha_{gxg^{-1}}$ is the coefficient of x in $g^{-1}ag$. Thus, a is a linear combination of $\gamma_1, \ldots, \gamma_s$ and these elements are a basis of $Z(kG)$ and so $r = s$.

The above proof cannot be adapted directly to the general case. The problem is that the center of $kG/\text{rad}(kG)$ is a subspace of a quotient and hard to study; instead we shall examine a quotient space of $kG/\text{rad}(kG)$. To see this, we shall first 'dualize' the above argument. For any algebra A, we let $[A, A]$ be the subspace of A spanned by all elements $ab - ba$, $a, b \in A$, and we shall work with the quotient space $A/[A, A]$ instead of $Z(A)$. If A is the direct sum of its subalgebras $A_1, \ldots, A_r$ then $[A, A] = [A_1, A_1] + \cdots + [A_r, A_r]$ is a vector space direct sum. If A is a matrix algebra then $[A, A]$ is a subspace of codimension one. In fact, $[M_n(k), M_n(k)]$ consists of all matrices of trace zero, as is well known and easy to prove (see the exercises at the end of this section). Thus, if $kG = A_1 + \cdots + A_r$ is a direct sum of matrix algebras, then $[kG, kG]$ is of codimension r. However, if $g_1, \ldots, g_s$ are elements, one from each conjugacy class, then the cosets of $[kG, kG]$ containing these elements form a basis for $kG/[kG, kG]$. We shall not prove this here as we shall do much more when we prove Theorem 2. This implies again that $r = s$.

Let us take up the general case now. Set $T = [kG, kG]$ and $S = T + \text{rad}(kG)$. It is immediate that $S/\text{rad } kG$ is just $[kG/\text{rad}(kG), kG/\text{rad}(kG)]$ and so S has codimension in kG equal to the number of matrix summands of $kG/\text{rad}(kG)$, that is, the number of simple kG-modules. We can assume now that the characteristic p of k divides the order of G (or easily adapt the following argument in the contrary case). Let $x_1, \ldots, x_s$ be representations of the s conjugacy classes of elements of order not divisible by p, that is, the p'-elements of G. Let $x_{s+1}, \ldots, x_r$ be representations of the remaining conjugacy classes of G. We shall prove that $x_1 + S, \ldots, x_s + S$ form a basis for kG/S and so the theorem will be proved. The key observation is the following.

Lemma 5 S consists of the elements a of kG such that $a^{p^i} \in T$ for some $i \geqslant 0$.

Let us postpone the proof of this result and first finish the proof of the theorem. We note that the elements $x_1 + T, \ldots, x_r + T$ are a basis for kG/T. Indeed, if x and y are conjugate elements of G, say $y = gxg^{-1}$, then $x - y = x - gxg^{-1} = g^{-1} \cdot gx - gx \cdot g^{-1} \in T$ so the elements $x_i + T$ do span kG/T as the elements of G span kG. In order to establish the linear independence of this spanning set we let φ_i, $1 \leqslant i \leqslant r$, be the linear functional on the vector space kG which has value one on each element of G conjugate to x_i and vanishes on all other group elements. In particular, φ_i is constant on each conjugacy class of G so if $g, h \in G$ then

$$\varphi_i(gh - hg) = \varphi_i(g(hg)g^{-1} - hg) = 0.$$

Hence $\varphi_i(T) = 0$. If $\sum \alpha_j(x_j + T) = 0$, that is, $\sum \alpha_j x_j \in T$, then

$$0 = \varphi_i\left(\sum \alpha_j x_j\right) = \alpha_i$$

for each i, $1 \leqslant i \leqslant r$, and the independence is proved.

Next, let $g \in G$ and express $g = ux$ where $u^{p^i} = 1$, for some $i \geqslant 0$, x is a p'-element and $ux = xu$; the structure of the cyclic group $\langle g \rangle$ gives us such a factorization. Hence, $g^{p^i} = x^{p^i}$ and so $(g - x)^{p^i} = 0$. Lemma 5 now yields that $g - x \in S$. We apply this to the elements $x_j, j > s$, of order divisible by p: each is congruent to a p'-element modulo S which is congruent to one of $x_1, \ldots,$ x_s modulo T. Hence, $x_1 + S, \ldots, x_s + S$ certainly span kG/S. Finally, suppose that $\alpha_1 x_1 + \cdots + \alpha_s x_s \in S$, each $\alpha_j \in k$. Lemma 5 now gives us that

$$\alpha_1^{p^i} x_1^{p^i} + \cdots + \alpha_s^{p^i} x_s^{p^i} = (\alpha_1 x_1 + \cdots + \alpha_s x_s)^{p^i} \in T$$

for some $i \geqslant 0$. But the elements $x_1^{p^i}, \ldots, x_s^{p^i}$ are in different conjugacy classes of G. (If the order of G is the product of a power of p and an integer m, with p not dividing m, choose integers a and b with $ap^i + bm = 1$. Since x_j is a p'-element we have $x_j^m = 1$ and so $(x_j^{p^i})^a = x_j$ and the integer a is independent of j.) Thus, they are linearly independent modulo T so each $\alpha_j^{p^i} = 0$, that is, $\alpha_j = 0$.

This leaves Lemma 5 to be demonstrated. Let S_0 be the set of elements a of kG such that $a^{p^i} \in T$ for some $i \geqslant 0$. Thus, in $\bar{A} = kG/\mathrm{rad}(kG)$, the image $\bar{a}$ of a is an element whose p^ith power lies in $[\bar{A}, \bar{A}]$. That is, each component of $\bar{a}$, in the decomposition of $\bar{A}$ into a direct sum of matrix algebras, has p^ith power an element of trace zero. But, in characteristic p, the trace of the pth power of a matrix is the pth power of the trace, since the trace is the sum of the eigenvalues. Hence, $\bar{a} \in [\bar{A}, \bar{A}]$, that is, $a \in T + \mathrm{rad}(kG) = S$. We have shown that $S_0 \subseteq S$. But $S_0 \supseteq T$ and also $S_0 \supseteq \mathrm{rad}\, kG$, as each element of $\mathrm{rad}\, kG$ is nilpotent, so if we can prove that S_0 is closed under addition then it follows that $S_0 \supseteq T + \mathrm{rad}\, kG$, which is S, and so $S_0 = S$ as claimed.

To do this we now prove two statements: (1) If $a, b \in kG$ then $(a + b)^p \equiv a^p + b^p$ (modulo T), that is, the difference lies in T; (2) $T^p \subseteq T$. The power $(a + b)^p$ is a sum of 2^p terms each of which is a product of p factors. Two of these terms are a^p and b^p and the remaining $2^p - 2$ terms fall into sets of p each, consisting of a product and its p cyclic rearrangements. In order to demonstrate (1), it is enough to show that the sum of p such rearrangements is in T. However, if $a_1, \ldots, a_p \in kG$, then

$$a_1 a_2 \cdots a_p + a_2 \cdots a_p a_1 = a_1 a_2 \cdots a_p + a_1 a_2 \cdots a_p$$
$$+ ((a_2 \cdots a_p)a_1 - a_1(a_2 \cdots a_p))$$
$$\equiv 2a_1 \cdots a_p \text{ (modulo } T)$$

so the sum of the p cyclic rearrangements of $a_j \cdots a_p$ lies in T. In view of (1), in order to prove (2), we need only show that if $a, b \in kG$ then $(ab - bc)^p \in T$. However,

$$(ab - ba)^p \equiv (ab)^p - (ba)^p \text{ (modulo } T)$$
$$= a(ba \cdots b) - (ba \cdots b)a \in T.$$

Finally, let $a, b \in S_0$. Since $T^p \subseteq T$, it follows that there is a positive integer n such that a^{p^n} and b^{p^n} lie in T. It follows that $(a + b)^{p^n} \equiv a^{p^n} + b^{p^n}$ (modulo T) so $a + b \in T$, and the theorem is proved in its entirety.

Exercises

1 With the notation of the proof of Theorem 1, show that $k\sigma$ is a nilpotent ideal if p divides $|G|$.

2 If G is a p-group then $\mathrm{rad}(kG) = \Delta(G)$.

3 With the notation for $\mathrm{SL}(2, p)$, show that V_n, $n \leqslant p$, has exactly $n + 1$ submodules as a $k\langle g\rangle$-module.

4 Let A be an $n \times n$ upper triangular matrix with ones on the main diagonal and all non-zero entries on the diagonal above it. Prove that the Jordan canonical form of A is a single $n \times n$ Jordan block for the eigenvalue 1. Use this to give another proof of the simplicity of the modules V_n, $n \leqslant p$.

5 If N is a normal subgroup of G then $\mathrm{rad}(kN) = kN \cap \mathrm{rad}(kG)$.

6 If N is a normal subgroup of G and T is a simple kN-module then there is a simple kG-module S with T a summand of S_N.

7 Show that $[M_n(k), M_n(k)]$ is the set of matrices of trace zero. (Hint: Let e_{ij}, $1 \leqslant i, j \leqslant n$, be the usual matrix units and show that each e_{ij} and $e_{ii} - e_{jj}$, $i \neq j$, is a 'bracket'.)

II
Projective modules

In the previous chapter, we studied semisimple modules and used them to see how the A-module A could be described in terms of semisimple modules. In fact, the radical and socle series of any module allow us to visualize it as made up of many semisimple layers, like a many-layered cake. In this chapter, we shall slice the cake! We shall study how modules can be decomposed into direct sums of indecomposable modules with special attention to the most important case of the A-module A. We shall be able to give complete information about certain group algebras, for cases that will be of key importance for later developments.

4 Indecomposable modules

We shall begin by studying the general properties of modules which do not have non-trivial direct sum decompositions, the indecomposable modules, and how arbitrary modules can be expressed as direct sums of these indecomposable modules. In the following sections we shall apply these ideas to the A-module A and to group algebras in particular.

Our first goal is to establish the basic characterization of indecomposable modules in terms of their endomorphism algebras. An algebra A is said to be *local* (terminology adapted from commutative algebra) if $A/\mathrm{rad}\ A$ is isomorphic with k.

Lemma 1 *The algebra A is local if, and only if, every element of A is nilpotent or invertible.*

Proof First, suppose that each element of A is either nilpotent or invertible, so that certainly the same holds for the algebra $A/\mathrm{rad}\ A$. However, this quotient is the direct sum of matrix algebras. If there is more than one matrix algebra then the unit element of one of them is neither nilpotent nor invertible so $A/\text{rad } A$ is a matrix algebra. Finally, if it is an $n \times n$ matrix algebra with $n > 1$ then it also has elements which are neither nilpotent nor invertible so we conclude that $A/\text{rad } A$ is just 1×1 matrices over k, that is, the algebra k as claimed.

Conversely, suppose that A is local. If a is an element of A and $a \in \text{rad } A$ then certainly a is nilpotent, while if $a \notin \text{rad } A$ then $a = \lambda 1 + r$ where $\lambda \in k$, $r \in \text{rad } A$. In this case the element

$$\lambda^{-1}(1 - \lambda^{-1}r + \lambda^{-2}r^2 - \cdots),$$

which makes sense as r is nilpotent, is readily seen to be inverse of a by multiplying it with $\lambda \cdot 1 + r$.

Theorem 2 *The A-module U is indecomposable if, and only if, $\text{End}(U)$ is local.*

Proof If $U = U_1 + U_2$ is a non-trivial direct sum let π be the projection of U on U_1 along U_2 so π is the identity on U_1 and is zero on U_2. In particular, π is a singular linear transformation and an element of $\text{End}(U)$ so it is a non-invertible element of $\text{End}(U)$. Moreover, $\pi^2 = \pi$ so π is not nilpotent either and so $\text{End}(U)$ is not local.

On the other hand, let $\rho \in \text{End}(U)$ with ρ neither invertible nor nilpotent. For each $\lambda \in k$ let U_λ be the generalized eigenspace of ρ for the eigenvalue λ, so U_λ consists of all $u \in U$ which are annihilated by a positive power of $\rho - \lambda I$. Since ρ is a linear transformation, we have that U_λ is non-zero for a finite number of λ in k, the eigenvalues of ρ, and U is the direct sum of the subspaces $U_\lambda, \lambda \in k$. However, each U_λ is also a submodule of U: if $u \in U_\lambda$ and $(\rho - \lambda I)^n u = 0$ then $(\rho - \lambda I)^n au = a(\rho - \lambda I)^n u = 0$ for any $a \in A$. Now ρ is nilpotent if, and only if, its only eigenvalue is zero and it is invertible if, and only if, zero is not an eigenvalue (that is, ρ has zero kernel). Hence, our assumption implies that zero is an eigenvalue of ρ and that ρ has a non-zero eigenvalue as well. In particular, U is a non-trivial direct sum.

The principal result in this section is the Krull–Schmidt theorem on uniqueness of decomposition.

Theorem 3 *If M is an A-module and*

$$M = U_1 + \cdots + U_r,$$
$$M = V_1 + \cdots + V_s,$$

are two decompositions into the direct sum of indecomposable modules then $r = s$ and, after suitable renumbering, $U_i \cong V_i$, for all i.

Proof We shall prove that after a renumbering we have $U_1 \cong V_1$ and a direct decomposition

$$M = U_1 + V_2 + \cdots + V_s.$$

It then follows that

$$M/U_1 \cong V_2 \oplus \cdots \oplus V_s$$

while our hypothesis gives

$$M/U_1 \cong U_2 \oplus \cdots \oplus U_r$$

so that the result is now immediate by an obvious induction.

Let π_i be the projection of M onto U_i, $1 \leqslant i \leqslant r$, along the sum of all the U_j, $j \neq i$, and let ρ_i, $1 \leqslant i \leqslant s$, be similarly defined for the second decomposition of M. Hence, $I = \rho_1 + \cdots + \rho_s$ and so $\pi_1 = \pi_1 I = \pi_1 \rho_1 + \cdots + \pi_1 \rho_s$. Since U_1 is indecomposable, the restriction of $\pi_1 \rho_j$ to U_1 is either nilpotent or invertible. But, if every $\pi_1 \rho_j$ is nilpotent, that is, lies in rad $\text{End}(U_1)$, then so is their sum which is the restriction of π_1 to U_1. However, that restriction is the identity map of U_1 so it follows that there is j, $1 \leqslant j \leqslant s$, with the restriction of $\pi_1 \rho_j$ to U_1 invertible. After a renumbering of the modules we may assume that it is $\pi_1 \rho_1$ that we have focused on.

We therefore are considering the following modules and homomorphisms:

$$U_1 \xrightarrow{\rho_1} V_1 \xrightarrow{\pi_1} U_1.$$

Let W be the image of U_1 under ρ_1 and K the kernel of π_1 restricted to V_1. Now π_1 maps W isomorphically onto U_1 since the restriction of $\pi_1 \rho_1$ to U_1 is an isomorphism. Hence, if $v \in V_1$ then $\pi_1(v) = \pi_1(w)$ for some $w \in W$. Thus, $\pi_1(v - w) = 0$ and $v = w + (v - w) \in W + K$ so $V_1 = W + K$. Moreover, $W \cap K = 0$ inasmuch as π_1 is zero on K and one-to-one on W. Hence, V_1 is the direct sum of W and K so $K = 0$ and π_1 maps isomorphically onto U_1.

Therefore, it remains only to prove that

$$U_1 \cap (V_2 + \cdots + V_s) = 0$$

since then

$$\dim(U_1 + V_2 + \cdots + V_s) = \dim U_1 + \dim(V_2 + \cdots + V_s)$$
$$= \dim V_1 + \dim V_2 + \cdots + \dim V_s$$
$$= \dim M$$

so $M = U_1 + (V_2 + \cdots + V_s)$ is a direct decomposition. However, ρ_1 is an isomorphism on U_1, as we have just seen, and it vanishes on each of $V_2, \ldots,$ V_s, so the proof is now complete.

Corollary 4 *If* M, U *and* V *are* A-*modules with* $M \oplus U \cong M \oplus V$ *then* $U \cong V$.

Proof If we express M, U and V as the direct sum of indecomposable modules then the isomorphism classes of indecomposable summands, as well as the multiplicity with which they occur, are the same for $M \oplus U$ and $M \oplus V$, by the Krull–Schmidt theorem. Hence, subtracting the multiplicities for the summands of M, the indecomposable summands of U and V, up to isomorphism and multiplicity, are the same for U and V so certainly U is isomorphic with V.

We shall now devote the rest of this section to a number of illustrative examples. First, suppose G is cyclic of order n: we claim that kG has exactly n isomorphism classes of indecomposable modules. Let g be a generator for G and let V be a kG-module. If T is linear transformation induced on V by g then $T^n = I$ inasmuch as $g^n = 1$. Therefore, any eigenvalue of T on V is an nth root of unity, that is, a root of the polynomial $x^n - 1$. Moreover, from Jordan canonical form we know that V is the direct sum of Jordan blocks for the eigenvalues of T. This means

$$V = V_1 + \cdots + V_r$$

is a direct sum where each V_i is a Jordan block; that is, V_i has a basis $v_1, \ldots,$ v_r where

$$T(v_1) = \lambda v_1 + v_2,$$
$$\vdots$$
$$T(v_{r-1}) = \lambda v_{r-1} + v_r,$$
$$T(v_r) = \lambda v_r,$$

and λ is an eigenvalue of T. Recall that V_i has only a one-dimensional space of eigenvectors for T, the space spanned by v_r, as is easily checked. Therefore, V_i is indecomposable as a kG-module, since each summand in a decomposition would have an eigenvector for T. Hence, each indecomposable kG-module is a Jordan block $J_r(\lambda)$ of dimension r for an nth root of unity λ.

Assume now that $V = V_i$. We assert that if we express $n = p^a e$, where p does not divide e, then λ is an eth root of unity and $r \leqslant p^a$. Indeed, $1 = \lambda^n = (\lambda^e)^{p^a}$ so $\lambda^e = 1$, as the taking of pth powers is a one-to-one function on k. Since p does not divide e, the polynomial $x^e - 1$ has exactly e roots; let them be $\lambda = \lambda_1, \lambda_2,$

..., λ_e. Therefore, $T^e - I = (T - \lambda_1 I) \cdots (T - \lambda_e I)$, so $T^e - I = (T - \lambda I)S$ where S is a linear transformation commuting with T and S is non-singular since none of $\lambda_2, \ldots, \lambda_e$ is an eigenvalue for T on V. Hence,

$$0 = T^n - I = (T^e - I)^{p^a} = (T - \lambda I)^{p^a} S^{p^a}$$

so $(T - \lambda I)^{p^a} = 0$. However, $(T - \lambda I)v_i = v_{i+1}$, $1 \leqslant i < r$ and $(T - \lambda I)v_r = 0$, so it follows that $r \leqslant p^a$. We have now shown that V has one of n possible structures since λ can be one of e roots of unity and $r \leqslant p^a$ while $n = p^a e$. Conversely, if λ is an eth root of unity and $r \leqslant p^a$, let V be a Jordan block for the linear transformation T, for eigenvalue λ, of dimension r. Therefore, $(T - \lambda I)^r = 0$ so $(T - \lambda I)^{p^a} = 0$ and $(T^e - I)^{p^a} = 0$, as above, giving $T^n = I$, so V is a module for kG in a natural way. Moreover, different Jordan blocks give non-isomorphic kG-modules since the eigenvalue for g and the dimension characterize the block.

Now let us turn to the structure of these indecomposable kG-modules. First, the simple kG-modules are exactly the one-dimensional Jordan blocks. Indeed, if $r > 1$, with the above notation, then the space spanned by $v_2, \ldots, v_r$ is a submodule, in fact, is a Jordan block of dimension $r - 1$ for λ. Hence, there are exactly e simple kG-modules, which is just as it should be since G has exactly e elements of order not divisible by p.

Next, we wish to determine the radical and socle series of V. In kG we have $(g^e - 1)^{p^a} = g^n - 1 = 0$ so $g^e - 1$ is a nilpotent element and therefore is in rad kG since kG is commutative. Now $(g^e - 1)V$ is the subspace W spanned by $v_2, \ldots, v_r$ since $(g - \lambda_2 1) \cdots (g - \lambda_e 1)V = V$ and $(g - \lambda 1)V$ is this subspace. Hence, W is contained in rad $V = (\text{rad } kG)V$ so it must be rad V as it has codimension one. Thus, rad V is the Jordan block for λ of dimension one less. Iterating this argument, we have that the radical series descends one dimension at a time. The successive quotients of this radical series are all the same simple kG-module, the one corresponding to λ. Similarly, the subspace spanned by v_r is the socle of V. Indeed, the socle is semisimple so is the sum of eigenspaces for g, but v_r spans the only eigenspace in V. The quotient of V by soc(V) has a basis consisting of the images of $v_1, \ldots, v_{r-1}$, so is also a Jordan block for λ. Thus, $V/\text{soc}(V)$ also has a simple socle so soc$^2(V)$ is spanned by v_r and v_{r-1}. Thus, each quotient of successive terms of the socle series of V is also the simple kG-module corresponding with λ.

This type of structure is an example of a more general phenomenon. The next result shows this as well as proving that the terms of the radical series of V are the only submodules of V.

Proposition 5 *If U is an A-module then the following are equivalent:*

(1) *U has a unique composition series;*

(2) *The successive quotients of the radical series of U are simple;*
(3) *The successive quotients of the socle series of U are simple.*

Proof Suppose that (1) holds and that

$$U = U_0 \supset U_1 \supset \cdots \supset U_n = 0$$

is the unique series. In particular, U_1 is the unique maximal submodule of U
since each such submodule is a term in a composition series. Hence, U_1 is
the intersection of all maximal submodules of U, that is, $U_1 = \operatorname{rad}(U)$. Now
U_1 must also have a unique composition series since U does so we also have
that $\operatorname{rad} U_1 = U_2$, that is, $U_2 = \operatorname{rad}^2(U)$. Hence, the given composition series
is the radical series and the successive quotients are simple.

On the other hand, suppose that (2) holds. Since $U/\operatorname{rad}(U)$ is simple this
means that $\operatorname{rad}(U)$ is the only maximal submodule of U and therefore any
composition series of U must have $U \supset \operatorname{rad}(U)$ as the first two terms.
Similarly, the radical quotient of $\operatorname{rad}(U)$ is simple so any composition series
of $\operatorname{rad} U$ begins $\operatorname{rad}(U) \supset \operatorname{rad}^2(U)$. Hence, any composition series of U
begins $U \supset \operatorname{rad}(U) \supset \operatorname{rad}^2(U)$ and, in general, the radical series of U is the
unique composition series of U. Thus, (1) and (2) are equivalent.

An entirely similar argument shows that (1) and (3) are equivalent. One
need only imitate the above arguments using the last terms of composition
series and the socle series of U.

A module satisfying the conditions of the proposition is called *uniserial*. We
have shown above that all indecomposable modules for cyclic groups are
uniserial.

We conclude this section by looking at a non-cyclic group: assume that G
is the direct product of two cyclic groups of order p with generators x and y.
We shall show that kG has infinitely many indecomposable modules by
constructing an infinite family of such modules.

For each positive integer n, let V_n be a $2n$-dimensional vector space over k
with basis $v_1, \ldots, v_n, w_1, \ldots, w_n$. Let X be the linear transformation of V
such that $Xv_i = w_i$, $Xw_i = 0$, $1 \leqslant i \leqslant n$. We can picture X as follows:

$$
\begin{array}{cccc}
v_1 & v_2 & \cdots & v_n \\
x\downarrow & x\downarrow & & x\downarrow \\
w_1 & w_2 & \cdots & w_n
\end{array}
$$

Here the arrows denote the images under X and an absence of an arrow
emanating from w_i indicates $Xw_i = 0$. Next let Y be the linear transforma-
tion such that $Yv_i = w_{i+1}$, $1 \leqslant i \leqslant n-1$, $Yv_n = 0$, $Yw_j = 0$, $1 \leqslant j \leqslant n$. We can

picture X and Y together as follows:

$$
\begin{array}{cccc}
v_1 & v_2 & \cdots & v_n \\
x\downarrow^{Y}\searrow & x\downarrow^{Y}\searrow & & {}^{Y}\searrow x\downarrow \\
w_1 & w_2 & \cdots & w_n
\end{array}
$$

Thus $X^2 = Y^2 = 0$ and $XY = YX = 0$. Since k has characteristic p this implies that $(I + X)^p = (I + Y)^p = I$. Since X and Y commute we can make V into a kG-module by setting $xv = (I + X)v$, $yv = (I + Y)v$ for all $v \in V$.

We shall prove that V_n is indecomposable so kG has infinitely many isomorphism classes of such modules. With regard to the given basis for V the linear transformations which correspond to the actions of x and y have the following matrices:

$$
\begin{pmatrix} I & 0 \\ I & I \end{pmatrix}, \quad \begin{pmatrix} I & 0 \\ N & I \end{pmatrix}
$$

where N is the $n \times n$ matrix,

$$
N = \begin{pmatrix} 0 & & & \\ 1 & 0 & & \\ \vdots & & \ddots & \\ 0 & \cdots & 1 & 0 \end{pmatrix},
$$

with all entries zero except just below the main diagonal, and I and 0 denote the $n \times n$ identity and zero matrices. The endomorphism algebra of V is isomorphic with the algebra of all $2n \times 2n$ matrices which commute with the above two $2n \times 2n$ matrices, since an endomorphism of V is just a linear transformation of V which commutes with the action of x and y.

However, let

$$
\begin{pmatrix} A & B \\ C & D \end{pmatrix}
$$

be a $2n \times 2n$ matrix, where A, B, C, D are $n \times n$ matrices. Thus,

$$
\begin{pmatrix} A & B \\ C & D \end{pmatrix}\begin{pmatrix} I & 0 \\ I & I \end{pmatrix} = \begin{pmatrix} A+B & B \\ C+D & D \end{pmatrix},
$$

$$
\begin{pmatrix} I & 0 \\ I & I \end{pmatrix}\begin{pmatrix} A & B \\ C & D \end{pmatrix} = \begin{pmatrix} A & B \\ A+C & B+D \end{pmatrix}.
$$

Hence, the condition for commutativity is $A = D$, $B = 0$. However,

$$
\begin{pmatrix} A & 0 \\ C & A \end{pmatrix}\begin{pmatrix} I & 0 \\ N & I \end{pmatrix} = \begin{pmatrix} A & 0 \\ C+AN & A \end{pmatrix},
$$

$$
\begin{pmatrix} I & 0 \\ N & I \end{pmatrix}\begin{pmatrix} A & 0 \\ C & A \end{pmatrix} = \begin{pmatrix} A & 0 \\ NA+C & A \end{pmatrix}
$$

so the matrices which commute with the two given ones are exactly those of the form

$$\begin{pmatrix} A & 0 \\ C & A \end{pmatrix}, \quad AN = NA.$$

However, it is also easy to see that A commutes with N if, and only if, it has the form

$$A = \begin{pmatrix} \lambda_1 & & & & \\ \lambda_2 & \lambda_1 & & & \\ \lambda_3 & \lambda_2 & \lambda_1 & & \\ \vdots & & & \ddots & \\ \lambda_n & & & & \lambda_1 \end{pmatrix}$$

that is, A is lower triangular with constant diagonals. The $2n \times 2n$ commuting matrices with zero main diagonal are thus of codimension one and they are, furthermore, a nilpotent ideal being the intersection with $\text{rad}(T_{2n}(k))$. Hence, the quotient of $\text{End}(V)$ by its radical is one-dimensional, so $\text{End}(V)$ is local and V is indeed indecomposable.

Exercises

We keep the notation of the example just calculated.

1 Let W be a two-dimensional vector space over k with basis u, v. If α, $\beta \in k$ show that V has a unique kG-module structure such that

$$xu = u + \alpha v, \quad xv = v,$$
$$yu = u + \beta v, \quad yv = v.$$

2 If $V_{\alpha,\beta}$ is the module just constructed then prove that $V_{\alpha,\beta}$ is indecomposable unless $(\alpha, \beta) = (0, 0)$.

3 Prove that $V_{(\alpha,\beta)}$ and $V_{(\gamma,\delta)}$ are isomorphic if, and only if, (α, β) and (γ, δ) are proportional.

4 Establish that any two-dimensional kG-module is isomorphic with some $V_{\alpha,\beta}$.

5 Prove that there are exactly two isomorphic classes of indecomposable three-dimensional kG-modules V with $(x-1)^2 V = (y-1)^2 V = 0$.

5 Free modules

We have already encountered the A-modules isomorphic with a direct sum $A \oplus \cdots \oplus A$. Such A-modules are called *free modules* and are the subject of this section. We begin with a very useful characterization of these modules but our main interest lies in their decomposition into direct sums of indecomposable modules. We are most interested in the structure of the summands of the A-module A in such a decomposition.

Proposition 1 *An A-module U is free if, and only if, U has a subspace X such that any linear transformation from X to any A-module V extends uniquely to a module homomorphism of U to V.*

Proof Let $U = A \oplus \cdots \oplus A$ be the direct sum of A n times. Let X be the subspace consisting of n-tuples whose entries are scalar multiples of the unit element of A. Let e_i, $1 \leqslant i \leqslant n$, be the n-tuple with all entries zero except for the ith which is 1, so $e_1, \ldots, e_n$ is a basis of X. Let φ be a linear transformation from X to V. If Φ is an extension to U which is a module homomorphism and $u = (a_1, \ldots, a_n) \in U$ then

$$\Phi(u) = \Phi(a_1 e_1 + \cdots + a_n e_n)$$
$$= \sum a_i \Phi(e_i)$$
$$= \sum a_i \varphi(e_i)$$

so Φ is uniquely determined. Moreover, it is quite easy to check that this formula for Φ does indeed give a module homomorphism extending φ.

On the other hand, suppose that W is an A-module possessing a subspace Y with the stated properties and say that Y has dimension n. We shall prove that $W \cong U$. Let φ be an isomorphism of X onto Y and let ψ be its inverse so $\varphi\psi$ and $\psi\varphi$ are the identity maps of Y and X, respectively. Let Φ and Ψ be the unique extensions of φ and ψ to module homomorphisms. It follows that $\Phi\Psi$ is a module homomorphism of W to W extending the identity linear transformation $\varphi\psi$ of Y. But the identity isomorphism of W also has this property so, by the uniqueness property of Y, $\Phi\Psi$ is the identity isomorphism of W. Similarly, $\Psi\Phi$ is the identity isomorphism of U so $U \cong W$ as claimed.

We can now give the fundamental characterization of summands of free modules.

Theorem 2 *If U is an A-module then the following are equivalent:*
   (1) *U is a direct summand of a free module;*
   (2) *If φ is a homomorphism of the A-module V onto U then the kernel of φ is a direct summand of V;*
   (3) *If φ is a homomorphism of the A-module V onto the A-module W and ψ is a homomorphism of U to W then there exists a homomorphism ρ of U to V such that $\varphi\rho = \psi$.*

The modules with these properties are called *projective*. Before proceeding with the proof we wish to make some comments. First, the last property is best understood in terms of the usual diagram:

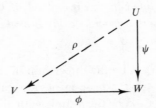

There is some standard terminology connected with the second property. If φ is a homomorphism of an A-module V onto any A-module U we say it *splits* if the kernel K is a direct summand. There is another useful formulation of this notion, which we shall now describe. Let $V = K + L$ be a direct sum so φ maps L onto U since V is mapped onto U and $\varphi(K) = 0$. Moreover, φ is one-to-one on L, inasmuch as $K \cap L = 0$. Hence, there is a homomorphism ψ of U to V which is the inverse of the restriction of φ to L. Thus $\varphi\psi$ is the identity map of U. Conversely, suppose that ψ is a homomorphism of U to V such that $\varphi\psi$ is the identity map of U. We claim that $V = K + \psi(U)$ is a direct sum so that φ splits. Indeed, $K \cap \psi(U) = 0$ since $\varphi(K) = 0$ and φ is one-to-one on $\psi(U)$ as $\varphi\psi$ is certainly one-to-one. Moreover, if $v \in V$ then

$$v = \psi(\varphi(v)) + v - \psi(\varphi(v))$$

and $\psi(\varphi(v)) \in \psi(U)$ while $v - \psi(\varphi(v)) \in K$ since its image under φ is $\varphi(v) - (\varphi\psi)(\varphi(v))$.

Proof We shall first prove that (1) implies (3). Let F be a free module with U as a direct summand, say $F = U + U'$ is a direct sum. Extend ψ to a homomorphism, also called ψ, of F to W which is zero in U'. It suffices to show there is a homomorphism ρ of F to V such that $\varphi\rho = \psi$, since its restriction to U will have the desired property. Let X be a subspace of F as described by Theorem 1. Since X is just a vector space there is certainly a linear transformation λ of X to V such that $\varphi\lambda$ is the restriction of ψ to X. Let ρ be the extension of λ to a homomorphism of U to V. Thus, $\varphi\rho$ and ψ are homomorphisms of U to W which agree on X. The uniqueness property of X implies that $\varphi\rho = \psi$.

Let φ now be as in (2) and assume that (3) is a property of U. We then have the diagram

$$U$$
$$\downarrow 1$$
$$V \xrightarrow{\varphi} U$$

so, by (3), there is a homomorphism ψ of U to V such that $\varphi\psi$ is the identity on U, that is, by the above discussion, φ splits, as required.

Finally, assume that (2) holds. There is a free module F and a homomorphism φ of F onto U, since this holds for any A-module. By (2), φ is split so U is isomorphic with a direct summand of F as claimed.

Now we are ready to discuss the important class of indecomposable projective modules. By the Krull–Schmidt theorem these are exactly the indecomposable summands of the free module A.

Theorem 3 *There is a one-to-one correspondence between isomorphism classes of indecomposable projective A-modules and isomorphism classes of simple A-modules given by associating $P/\mathrm{rad}\,P$ to each indecomposable projective A-module P.*

Thus, we are saying a number of things. First, $P/\mathrm{rad}\,P$ is simple, that each simple module so arises and finally, if P and Q are both indecomposable projective A-modules and $P/\mathrm{rad}(P) \cong Q/\mathrm{rad}(Q)$ then $P \cong Q$. Note that the property of having the quotient by the radical being simple is the same as having a unique maximal submodule. Indeed, if U is any A-module and U has a unique maximal submodule M then $M = \mathrm{rad}(U)$, as rad U is the intersection of all maximal submodules of U, so $U/\mathrm{rad}(U)$ is simple. Conversely, if $U/\mathrm{rad}(U)$ is simple then $\mathrm{rad}(U)$ is maximal so that if M were another maximal submodule we would have

$$\mathrm{rad}(U) \subseteq (\mathrm{rad}(U)) \cap M \subset \mathrm{rad}(U).$$

Before proving this theorem, let us first observe a consequence.

Corollary 4 *In a decomposition of the free module A into a direct sum of indecomposable submodules, each isomorphism type of indecomposable projective module occurs as many times as the dimension of the corresponding simple module.*

Indeed, suppose that $A = P_1 + \cdots + P_r$ is such a decomposition. Multiplying through by rad A we have that rad $A = \mathrm{rad}(P_1) + \cdots + \mathrm{rad}(P_r)$ so

$$A/\mathrm{rad}(A) \cong P_1/\mathrm{rad}(P_1) + \cdots + P_r/\mathrm{rad}(P_r).$$

However, this is now just a decomposition of $A/\mathrm{rad}\ A$ into a direct sum of simple modules. Hence, each isomorphism class of simple modules occurs as many times as its dimension, by our analysis of semisimple algebras. Theorem 3 now implies the corollary.

Let us now prove Theorem 3. First, we show that if P is a projective A-module then the algebra $\mathrm{End}(P/\mathrm{rad}(P))$ is a homomorphic image of the algebra $\mathrm{End}(P)$. Indeed, if φ is an endomorphism of P then $\varphi(\mathrm{rad}(P)) \subseteq \mathrm{rad}(P)$ as $\mathrm{rad}(P) = (\mathrm{rad}\ A)P$. Hence, there is an endomorphism $\bar{\varphi}$ of $P/\mathrm{rad}(P)$ such that $\bar{\varphi}(x + \mathrm{rad}(P)) = \varphi(x) + \mathrm{rad}\ P$, for any $x \in P$, and the map that sends φ to $\bar{\varphi}$ is an algebra homomorphism, as is also easily checked. If ρ is the natural map of P onto $P/\mathrm{rad}(P)$ then we can reformulate this as follows: $\bar{\varphi}$ is the unique homomorphism of $P/\mathrm{rad}(P)$ to $P/\mathrm{rad}(P)$ such that the following diagram commutes (that is, $\rho\varphi = \bar{\varphi}\rho$):

$$
\begin{array}{ccc}
P & \xrightarrow{\ \varphi\ } & P \\
{\scriptstyle\rho}\downarrow & & \downarrow{\scriptstyle\rho} \\
P/\mathrm{rad}(P) & \xrightarrow[\bar{\varphi}]{} & P/\mathrm{rad}(P)
\end{array}
$$

Hence, to show that the map that sends φ to $\bar{\varphi}$ has $\mathrm{End}(P/\mathrm{rad}(P))$ as its image, it is enough to prove that there is such a φ, given $\bar{\varphi}$. However, given $\bar{\varphi}$ we have that $\bar{\varphi}\rho$ is a homomorphism of the projective module P to the module $P/\mathrm{rad}(P)$ and ρ is a homomorphism of the module P onto $P/\mathrm{rad}(P)$, so, by the characterization of projective modules, there is a homomorphism φ of P to P such that $\rho\varphi = \bar{\varphi}\rho$, as required.

Now, if P is indecomposable then $\mathrm{End}(P)$ is local, each element of $\mathrm{End}(P)$ is invertible or nilpotent and so the image $\mathrm{End}(P/\mathrm{rad}(P))$ also has this property and is local. But $\mathrm{End}(P/\mathrm{rad}(P))$ is semisimple, as the endomorphism algebra of a semisimple module is the direct sum of matrix algebras. However, this algebra is also local, that is, one-dimensional modulo its radical, so we deduce that $\mathrm{End}(P/\mathrm{rad}(P)) \cong k$. Thus, $P/\mathrm{rad}\ P$ is simple as any non-simple semisimple module has a larger endomorphism algebra.

Next, suppose that P and Q are indecomposable projective A-modules and that $P/\mathrm{rad}(P) \cong Q/\mathrm{rad}(Q)$. Hence, we have a commutative diagram

$$
\begin{array}{ccc}
P & \to & Q \\
\downarrow & & \downarrow \\
P/\mathrm{rad}(P) & \to & Q/\mathrm{rad}(Q)
\end{array}
$$

where the vertical maps are the natural maps and the bottom horizontal map is an isomorphism. Hence, the image of P in Q cannot be contained in rad(Q): for then the composition of the map from P to Q with the natural map of Q to $Q/\text{rad}(Q)$ would be zero, while the other composition is certainly not zero. But $Q/\text{rad}(Q)$ is simple and rad(Q) is the intersection of all maximal submodules of Q so that it must be the unique maximal submodule of Q. Since the image of P is not contained in rad(Q), it must be all of Q. Since Q is projective, it follows that Q is isomorphic with a direct summand of P and so $P \cong Q$, as P is indecomposable.

Finally, suppose that S is a simple A-module. Let F be a free module such that S is a homomorphic image of F. Hence, there is an indecomposable summand P of F which has a non-zero homomorphism to S. It must be onto, as S is simple, so S is a homomorphic image of $P/\text{rad}(P)$, that is, $S \cong P/\text{rad}(P)$. The theorem is now fully proved.

The next-to-last paragraph applies to a module Q which is not projective and shows that Q is a homomorphic image of P. We restate this useful fact.

Lemma 5 *If U is an A-module, $U/\text{rad}(U)$ is simple and isomorphic with $P/\text{rad}(P)$, where P is an indecomposable projective module, then U is a homomorphic image of P.*

We now devote the rest of this section to group algebras. We begin with the 'real meaning' of Lagrange's theorem.

Theorem 6 *If P is a projective kG-module and H is a subgroup of G then P_H is a projective kH-module.*

Proof It suffices to prove that the free kG-module kG is free as a kH-module, since every projective module is a direct summand of a direct sum of modules each isomorphic with kG. Let $C_1, \ldots, C_r$ be the cosets of H in G so G is their disjoint union and if $x_i \in C_i$ then $C_i = Hx_i$. Let kC_i be the vector space in kG consisting of all linear combinations of elements of C_i. Clearly, kC_i is a kH-module. Moreover, as such, $kC_i \cong kH$ since the linear map which sends each h in H to hx_i is a module isomorphism. Hence, $(kG)_H$ is certainly free, $(kG)_H \cong kH \oplus \cdots \oplus kH$, the number of times being the index r of H in G.

Corollary 7 *If a Sylow p-subgroup P of the group G has order p^a then every projective kG-module has dimension divisible by p^a.*

Proof Since kP has a unique simple module and it is of dimension one, Corollary 4 implies that kP is indecomposable. Hence, every projective kP-

module is free and certainly of dimension divisible by p^a. Theorem 6 now gives this result immediately.

Let us now turn to our first example and assume that $G = \langle g \rangle$ is cyclic of order $n = p^a e$, as before, so p does not divide e. We have seen that to each eth root of unity there is a one-dimensional simple kG-module, call it S_λ, in which g acts by multiplication by λ. We also saw that to each integer m, $1 \leqslant m \leqslant p^a$, there is a uniserial module of dimension m, with each composition factor isomorphic with S_λ, and that, as λ varies over the e eth roots of unity and m varies over its range, the $n = ep^a$ uniserial modules give us all the indecomposable kG-modules. Corollary 7 now implies that the p^a-dimensional uniserial module with all composition factors isomorphic with S_λ is the indecomposable projective corresponding to S_λ: no other indecomposable module has S_λ as a quotient module and dimension divisible by p^a. In particular, we now have the structure of kG as a kG-module. Since the simple modules are of dimension one, kG is isomorphic with the direct sum of each of the e uniserial modules of dimension p^a. This can be pictured as follows:

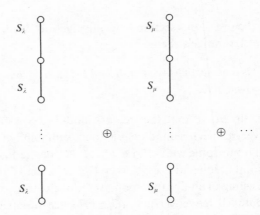

Next we will prove a lemma for our arbitrary group G. This will be useful in dealing with the special case that will conclude this section.

Lemma 8 *If R is a normal Sylow p-subgroup of G and U is a kG-module then* $\operatorname{rad}(U) = \operatorname{rad}(U_R)$. *Moreover, if R is cyclic with generator x then* $\operatorname{rad}(U) = (1-x)U$.

Proof If $g \in G$ then g induces an automorphism of the algebra kR by conjugation; in particular, $g \operatorname{rad}(kR)g^{-1} = \operatorname{rad}(kR)$. Therefore,

$$g \operatorname{rad}(U_R) = g \operatorname{rad}(kR) \cdot U$$
$$= g \operatorname{rad}(kR)g^{-1} \cdot gU$$
$$= \operatorname{rad}(kR)U$$
$$= \operatorname{rad}(U_R)$$

so $\operatorname{rad}(U_R)$ is a kG-submodule of U. The quotient $U/\operatorname{rad}(U_R)$ is a semisimple kR-module and since R is a p-group this means each element of R induces the identity linear transformation on $U/\operatorname{rad}(U_R)$. Hence, $U/\operatorname{rad}(U_R)$ is a module for G/R, which is of order not divisible by p as R is a Sylow subgroup. Hence, $U/\operatorname{rad}(U_R)$ is a semisimple module and so $\operatorname{rad}(U_R) \supseteq \operatorname{rad}(U)$. But $U/\operatorname{rad}(U)$ is a semisimple kG-module so is semisimple as a kR-module, by Clifford's theorem, and therefore $\operatorname{rad}(U) \subseteq \operatorname{rad}(U_R)$. This proves the first assertion.

If x generates R then $(1-x)U$ is a kR-submodule since R is certainly abelian. Moreover, $U/(1-x)U$ is semisimple, since x, and hence each element of R, induces the identity linear transformation on this quotient module. Thus, $(1-x)U \supseteq \operatorname{rad}(U_R)$. But $1-x$ annihilates the trivial kR-module, which is the only simple kR-module, so $1-x \in \operatorname{rad}(kR)$ and $(1-x)U \subseteq \operatorname{rad}(kR)U = \operatorname{rad}(U_R)$ and the lemma is proven.

We shall now study this case further: assume that R is a cyclic normal Sylow p-subgroup of G of order p^a with generator x. Our aim is to completely describe the indecomposable kG-modules and see how the answer is very much like the case for cyclic groups. We shall do this in this section for the indecomposable projective modules and finish this task in the next section with the use of some further methods.

Let P be the indecomposable projective module corresponding to the simple module S. If S is of dimension d then we assert that $P_R \cong kR \oplus \cdots \oplus kR$, d times, P has radical length p^a and each successive quotient of the radical series of P is also d-dimensional. Now P_R is a free module since it is projective and every kR-projective module is free. Hence, P_R is a direct sum of modules each isomorphic with kR. Now $P/\operatorname{rad}(P)$ is d-dimensional so $P_R/\operatorname{rad}(P_R)$ is also, as $\operatorname{rad}(P) = \operatorname{rad}(P_R)$ by Lemma 8. But $kR/\operatorname{rad}(kR)$ is one-dimensional, as kR is the indecomposable projective corresponding to the trivial module, so we must have that P_R is the direct sum of exactly d copies of kR. Moreover, by Lemma 8, the radical series of P and P_R coincide. Since kR is uniserial of composition length p^a and each successive quotient of the radical series of kR is one-dimensional, the other assertions about P are immediate.

Suppose that $S = k$, the trivial module, so $P/\mathrm{rad}(P) \cong k$ and $\mathrm{rad}(P)/\mathrm{rad}^2(P) \equiv W$ is also a one-dimensional simple module. If we set $M = P/\mathrm{rad}^2(P)$ then we have an exact sequence

$$0 \to W \to M \to k \to 0$$

which is not split, that is, W is not a direct summand of M. This two-dimensional uniserial module M is very important for us.

However, in order to use M we have to introduce another general idea (which works for all groups G). If U and V are kG-modules then the vector space $U \otimes V$, the tensor product of U and V, is also a kG-module in a natural way: if $g \in G$, $u \in U$, $v \in V$, then set $g(u \otimes v) = gu \otimes gv$. This gives a well-defined action of g on $U \otimes V$, because of the bilinearity of the definition. Moreover, it is easy to check that now $U \otimes V$ is a module. A number of properties of tensor products of vector spaces carry over to modules. For example, if W is another kG-module then $U \otimes (V \otimes W) \cong (U \otimes V) \otimes W$. Furthermore, if V_1 is a submodule of V then $U \otimes V_1$ is a submodule of $U \otimes V$ and the quotient of $U \otimes V$ by $U \otimes V_1$ is isomorphic with $U \otimes V/V_1$. These properties can be verified directly.

Returning to the situation we are investigating, consider the tensor product $S \otimes M$. It has a submodule $S \otimes W$ and the quotient by this submodule is isomorphic with $S \otimes k$, which is obviously isomorphic with S. We claim that $S \otimes W$ is also simple and that $S \otimes M$ is not semisimple. To prove the first statement we construct another kG-module. Let W_1 be a one-dimensional vector space over k and let $g \in G$ act on W_1 by multiplication by the scalar μ^{-1}, where μ is the scalar given by the action of g on W. It is trivial to verify that W_1 is a kG-module and that $W \otimes W_1 \cong k$. Thus,

$$(S \otimes W) \otimes W_1 \cong S \otimes (W \otimes W_1) \cong S$$

so if $S \otimes W$ had a non-zero proper submodule so would $(S \otimes W) \otimes W_1$, contradicting this simplicity of S. Let us now establish the second statement. By Lemma 8, we have $(1 - x)M \neq 0$ so choose $0 \neq m \in M$ with $(1 - x)m \neq 0$. Let s be a non-zero element of S; again by Lemma 8, in order to prove that $S \otimes M$ is not semisimple, it suffices to show that $(1 - x)(x \otimes m) \neq 0$. But $x \in R$, a normal p-subgroup, S_R is semisimple so $xs = s$. Hence,

$$
\begin{aligned}
(x - 1)(s \otimes m) &= x(s \otimes m) - s \otimes m \\
&= xs \otimes xm - s \otimes m \\
&= s \otimes xm - s \otimes m \\
&= s \otimes (x - 1)m \neq 0
\end{aligned}
$$

as $s \neq 0$ and $(x - 1)m \neq 0$. We now have that $S \otimes M$ is uniserial of length two. Indeed, $S \otimes W$ is a submodule and the quotient of $S \otimes M$ by it is simple so

$S \otimes W \supseteq \mathrm{rad}(S \otimes M)$. But $S \otimes W$ is simple and $\mathrm{rad}(S \otimes M) \neq 0$ so $S \otimes W = \mathrm{rad}(S \otimes M)$.

We return now to the projective module P with $P/\mathrm{rad}(P) \cong S$. We claim that $\mathrm{rad}(P)/\mathrm{rad}^2(P) \cong S \otimes W$. Indeed, since $S \otimes M/\mathrm{rad}(S \otimes M) \cong S$ we have that $S \otimes M$ is a homomorphic image of P. Hence,

$$\mathrm{rad}(S \otimes M)/\mathrm{rad}^2(S \otimes M) = S \otimes W$$

is a homomorphic image of $\mathrm{rad}(P)/\mathrm{rad}^2(P)$. But the latter quotient is also d-dimensional, as is $S \otimes W$, so our claim is valid.

Now suppose U is any module with $U/\mathrm{rad}\, U \cong S$. Again U is a homomorphic image of P so $\mathrm{rad}(U)/\mathrm{rad}^2(U)$ is either zero, so $U \cong S$, or it is isomorphic with $S \otimes W$, since it is a homomorphic image of $S \otimes W$. Of course, this holds for any simple module S; the module W is independent of our choice of S.

Finally, let us describe the radical series of P in detail. We know that it is of length p^a, that $P/\mathrm{rad}\, P \cong S$ and $\mathrm{rad}(P)/\mathrm{rad}^2(P) \cong S \otimes W$. But now $\mathrm{rad}(P)$ is a module, the quotient by its radical is the simple module $S \otimes W$ so we can apply the previous paragraph to the module $U = \mathrm{rad}(P)$ and simple module $S \otimes W$. We deduce that $\mathrm{rad}^2(P)/\mathrm{rad}^3(P) \cong S \otimes W \otimes W$ (provided $p^a \geqslant 3$). Continuing in this way, the next quotient of successive terms of the radical series is $S \otimes W \otimes W \otimes W$. Each of these quotients is simple, being the tensor product of a simple module and W. We therefore have that P is uniserial and we have a specific description of the simple factors involved in P.

Exercises

1 The indecomposable projective $T_n(k)$-module associated with the simple $T_n(k)$-module S_i, $1 \leqslant i \leqslant n$, is uniserial and the successive quotients of its radical series are $S_i, S_{i+1}, \ldots, S_n$.

2 If S is a simple A-module with corresponding projective module P then the multiplicity of S as a composition factor of the A-module U equals the dimension of $\mathrm{Hom}_A(P, U)$.

3 Determine the module W, of the end of the section, by directly constructing a module isomorphic with M in the following way. (Some knowledge of group theory is necessary.) Let $G = HR$, a semi-direct product, by the theorem of Schur–Zassenhaus. If $h \in H$ let $a(h)$ be an integer such that $hxh^{-1} = x^{a(h)}$. Let $\alpha(h) = a(h) \cdot 1 \in k$. Show that there is a two-dimensional module with basis u, v such that

$$xu = u + v, \quad xv = v,$$
$$hu = u, \quad hv = \alpha(h)v.$$

Prove that it is isomorphic with M. Show that in W the element $g = hx_i$, $h \in H$, acts by multiplication by $\alpha(h)$.

4 Let φ be a one-to-one homomorphism of the A-module U into the A-module V. Show that $\varphi(U)$ is a direct summand of V if, and only if, there is a homomorphism ψ of V to U such that $\psi\varphi$ is the identity map of U.

5 Let R be cyclic of order p^a with generator x. Set $X = x - 1 \in kR$. Show that $1, X, X^2, \ldots, X^{p^a-1}$ is a basis of kR and that $X, X^2, \ldots, X^{p^a-1}$ is a basis for $\text{rad}(kR)$.

6 Duality

The notion of duality is basic in linear algebra and we shall generalize this idea, from vector spaces, that is, k-modules, to kG-modules. This will be a very powerful tool and we shall use it immediately to get deeper information about projective modules and to proceed further with our study of modules in the case of a normal cyclic Sylow p-subgroup.

Recall that if V is a vector space over k then the dual space V^* is just $\text{Hom}_k(V, k)$, the space of linear functionals. If V is a kG-module then V^* is also a kG-module by means of a natural construction. If $g \in G$ and $\varphi \in V^*$ then define $g\varphi$ to be the element of V^* such that $(g\varphi)(v) = \varphi(g^{-1}v)$ for any $v \in V$. It is indeed trivial to verify that $g\varphi \in V^*$ and that V^* is a kG-module. We leave these details to the reader, including the calculation that establishes $g_1(g_2\varphi) = (g_1g_2)\varphi$, whenever $g_1, g_2 \in G$. It should be observed how important is the use of the inverse of g, in the defining formula; at this point this construction depends on having a group algebra.

If ρ is a linear transformation from a vector space V to a vector space W then there is a corresponding linear transformation ρ^* (usually called the transpose) from W^* to V^*: if $\psi \in W^*$ then $\rho^*(\psi)$ is the element of V^* defined by $(\rho^*(\psi))v = \psi(\rho(v))$, for any v in V. Again, if V and W are kG-modules and ρ is a kG-homomorphism from V to W then ρ^* also has the desired property: it is a kG-homomorphism from the kG-module W^* to the kG-module V^*. This is also easy to establish, but here we shall present part of the direct argument. Let $\psi \in W^*$, $g \in G$; we claim that $g(\rho^*(\psi)) = \rho^*(g\psi)$. In fact, let $v \in V$, so

$$
\begin{aligned}
(g \cdot \rho^*(\psi))v &= (\rho^*(\psi))(g^{-1}v) \\
&= \psi(\rho(g^{-1}v)) \\
&= \psi(g^{-1}\rho(v)) \\
&= (g\psi)(\rho(v)) \\
&= (\rho^*(g\psi))v.
\end{aligned}
$$

Indeed, all the usual properties of duality carry over to kG-modules. For example, if V is a kG-module then $V \cong V^{**}$; the usual natural isomorphism of the vector space V onto the double dual V^{**} is also a kG-homomorphism. This, and other routine verifications, are left to the reader to carry out. The relations between subspaces, quotient spaces and duality also extend immediately: if W is a submodule of the kG-module V then V^* has a submodule naturally isomorphic with $(V/W)^*$ and the quotient of V^* by this submodule is naturally isomorphic with W^*. This is easily pictured as follows.

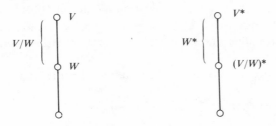

The usual argument carries right over. The subspace of W^* consisting of the linear functionals which vanish on W can be identified with the linear functionals on V/W. The quotient of V^* by this subspace is seen to be isomorphic with W^* by means of the map which sends an element of V^* to its restriction to W which is in W^*. At each step, the group operations pose no difficulties. It is also true that if V and W are kG-modules then $(V \oplus W)^* \cong V^* \oplus W^*$ and $(V \otimes W)^* \cong V^* \otimes W^*$. The first isomorphism follows from the map of $(V \oplus W)^*$ to $V^* \oplus W^*$ which sends a linear functional on $V + W$ to the pair consisting of its restrictions to V and W, respectively. The second isomorphism occurs because of a map from $V^* \otimes W^*$ to $(V \otimes W)^*$ which sends $\varphi \otimes \psi$, where $\varphi \in V^*$ and $\psi \in W^*$, to the linear functional on $V \otimes W$ which maps $v \otimes w$, $v \in V$, $w \in W$, to $\varphi(v)\psi(w)$.

Now let us turn to the connection between duality and the notions around the idea of semisimplicity. We need only one easy result.

Lemma 1 *The kG-module U is simple if, and only if, U^* is simple.*

Proof If U is not simple and has a non-zero proper submodule V then U^* has a non-zero submodule isomorphic with $(U/V)^*$ and is therefore not simple. Similarly, if U^* is not simple then U^{**} is not either, that is, U is not simple.

We therefore have a permutation of order two on the isomorphism classes of simple kG-modules. (This is one-to-one as $U_1^* \cong U_2^*$ yields $U_1 \cong U_2$ by the taking of double duals.) It is also a consequence that the kG-module U is semisimple if, and only if, U^* is semisimple. In fact, if U is the direct sum of simple submodules then U^* is also a direct sum, the sum of the duals of the simple submodules which are again simple. Similarly, if U^* is semisimple then so is $U^{**} \cong U$. This in turn gives us information connecting radicals, socles and duality. If U is any kG-module then $\text{soc}(U^*) \cong (U/\text{rad}(U))^*$ and $U^*/\text{rad}(U^*) \cong \text{soc}(U)^*$. Indeed, since $U/\text{rad}(U)$ is a quotient module of U, we have that U^* has a submodule isomorphic with $(U/\text{rad}(U))^*$ which is semisimple. On the other hand, each submodule of U^* is isomorphic with the dual of a quotient module of U. Hence, the first of these two isomorphism holds. The second can be established in an entirely similar manner.

Our next step is to bring in the notation of projectivity.

Lemma 2 *The dual of a free module is free.*

Of course, as usual, the converse holds by the use of double duals. By our discussion of direct sum, it suffices to see that $(kG)^* \cong kG$ in order to establish this lemma. For each $g \in G$ let $\varphi_g \in (kG)^*$ be the linear functional which has value 1 on g and vanishes on all other elements of G. There is a one-to-one linear transformation of the vector space kG onto the vector space $(kG)^*$ which sends each $g \in G$ to $\varphi_g \in (kG)^*$. This is a kG-homomorphism as well. If $g, h \in G$ then $h\varphi_g = \varphi_{hg}$ since, for any $x \in G$, we have

$$(h\varphi_g)(x) = \varphi_g(h^{-1}x)$$

so $h\varphi_g$ has value 1 on x if $g = h^{-1}x$, that is, $hg = x$, while $h\varphi_g$ vanishes on x if $g \neq h^{-1}x$, that is $hg \neq x$.

As a consequence of this we have that if a module P is projective then so is its dual P^*, inasmuch as P^* is a summand of the dual of any free module which has P as a summand. However, we know projective modules have certain characteristic properties and their duals have different properties! This gives us new information about projective modules.

Proposition 3 *If U is a kG-module then the following are equivalent:*
  (1) *U is a direct summand of a free module;*
  (2) *If φ is a one-to-one homomorphism of U into the kG-module V then $\varphi(U)$ is a direct summand of V;*
  (3) *If ψ is a one-to-one homomorphism of the kG-module W into the*

kG-module V and φ is a homomorphism of W to U then there is a homomorphism ρ of V to U such that φ = ρψ.

The last statement can be pictured as follows:

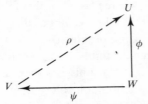

The modules with these properties are called *injective* so, in view of property (1), we will have the following theorem as a consequence as soon as we prove the proposition.

Theorem 4 *A kG-module is projective if, and only if, it is injective.*

Proof Suppose that U has the first property and let φ be as in (2). It follows that φ^* is a homomorphism of V^* onto U^*. However, U is projective, since (1) holds, so U^* is also projective and therefore, by the second property characterizing projective modules, there is a homomorphism of U^* to V^* whose composition with φ^* is the identity map of U^*. That is, there is a homomorphism ψ of V to U such that $\varphi^*\psi^*$ is the identity on U^* which means $\psi\varphi$ is the identity on U. But φ is one-to-one so V is the direct sum of $\varphi(U)$ and the kernel of ψ, as is easy to verify. On the other hand, if (2) holds then using duality it is easy to prove that U^* has the second property characterizing projective modules so U^* satisfies (1) and therefore, so does $U \cong (U^*)^*$. Similarly, it is easy, using duality, to verify that (1) and (3) are equivalent using the third property characterizing projective modules.

We now get as a consequence a fundamental property of projective modules for group algebras.

Corollary 5 *If P is an indecomposable projective kG-module than soc(P) is simple.*

Thus, the 'bottom' of P, as well as the 'top', is simple. Since P is indecomposable so is P^* and since P is projective we have that P^* is also. Hence $P^*/\mathrm{rad}(P^*)$ is simple. However, $\mathrm{soc}(P)^* \cong P^*/\mathrm{rad}(P^*)$ so $\mathrm{soc}(P)$, the dual of $\mathrm{soc}(P)^*$, is also simple. The last task of this section will be to establish the precise relation between $P/\mathrm{rad}(P)$ and $\mathrm{soc}(P)$ using a refinement of duality, but we shall first apply our results to a special type of group which we have already studied.

We assume now that G has a cyclic normal Sylow p-subgroup. We proved that each indecomposable projective kG-module was uniserial and of composition length p^a, where p^a is the order of the Sylow p-subgroup of G. We shall now establish that all indecomposable kG-modules are uniserial. This implies that an indecomposable kG-module U is characterized by its quotient $U/\mathrm{rad}(U)$, which is simple, and its composition length, which is between 1 and p^a. Indeed, if U is uniserial then certainly $U/\mathrm{rad}(U) \cong S$ is a simple module so U is a homomorphic image of the indecomposable projective module P which corresponds to S. But P has a unique quotient of each composition length between 1 and p^a and U is characterized as claimed. Moreover, if kG has exactly e simple modules this argument shows that kG has exactly ep^a indecomposable modules.

Therefore, let M be any kG-module. Choose a submodule U of M and a quotient module U/V of U such that U/V is uniserial and such that, among all uniserial quotients of submodules of M, the maximum composition length is achieved by U/V. We shall show that $U = U_0 + V$ is a direct sum, so $U_0 \cong U/V$, and that $M = U_0 + V_0$ is also a direct sum. If M is indecomposable this yields $M = U_0$ and M is indecomposable and all our assertions about kG-modules are established.

First, let $W/V = \mathrm{rad}(U/V)$ so $U/W \cong S$, a simple kG-module. Let P be an indecomposable projective module corresponding to S so there is a homomorphism ψ of P onto U/W. Since P is projective there is a homomorphism φ of P to U such that the composition of φ and the natural map of U onto U/W equals ψ. We set $U_0 = \varphi(P)$ so that $U_0 + W = U$. This implies that $U_0 + V = U$: $U_0 + V/V$ is a submodule of the uniserial module U/V and its sum with the unique maximal submodule W/V is U/V. Moreover,

U_0 is uniserial since it is a homomorphic image of P which is uniserial. If $U_0 \cap V \neq 0$ then U_0 has greater composition length than U/V, as $U_0/U_0 \cap V \cong U/V$, and this contradicts our choice of U and V. Hence, $U_0 \cap V = 0$ and we have established the first direct decomposition $U = U_0 + V$.

To get the second direct decomposition we use a 'dual' argument. Let $T \cong$

soc(U_0) so T is a simple module. Let Q be the indecomposable injective module with $T \cong$ soc(Q); this exists as one can simply take the dual of the indecomposable projective module corresponding to the simple module T^*. Since soc(U_0) is a submodule of U_0 there is a homomorphism φ of M to Q extending an isomorphism of soc(U_0) onto soc(Q); this exists by the third property characterizing injective modules in Proposition 3. Let V_0 be the kernel of φ so M/V_0 is uniserial since M/V_0 is isomorphic with a submodule of Q and Q is uniserial being indecomposable and projective. Moreover, $U_0 \cap V_0 = 0$; for if $U_0 \cap V_0 \neq 0$ it would have to contain soc(U_0), which is the unique simple submodule of U_0, and so we would have $\varphi(\text{soc}(U_0)) = 0$. But now

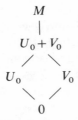

$U_0 + V_0/V_0$ is a uniserial module, isomorphic with U_0, contained in the uniserial module M/V_0, so, by our choice of U and V, we must have $M = U_0 + V_0$ and the second decomposition is demonstrated.

In dealing with duality, it is often possible to identify a vector space and its dual by means of a bilinear form. We shall use this idea for kG-modules; it is really prefaced by the isomorphism $kG \cong (KG)^*$. Let $(\ ,\ )$ be the bilinear form on the vector space kG such that if $g, h \in G$ then (g, h) is zero unless $gh = 1$, in which case the value is 1. This is a 'symmetric' form, in that $(a, b) = (b, a)$ whenever $a, b \in kG$, since this certainly holds for group elements. The form is non-degenerate, that is, if $(a, b) = 0$ for some $a \in kG$ and all $b \in kG$ then $a = 0$; indeed, the determinant of the form with respect to the basis of group elements is clearly ± 1. Finally, the form is 'associative', that is, $(ab, c) = (a, bc)$ whenever $a, b, c \in kG$, since this holds for group elements. We can use this form to establish an important result:

Theorem 6 *If P is an indecomposable projective kG-module then $P/\text{rad}(P) \cong \text{soc}(P)$.*

Proof Let $P/\text{rad}(P) \cong S$, a simple kG-module, and assume that soc(P) $\not\cong S$. Express $kG = Q + R$ as a direct sum of two submodules, where each indecomposable summand of Q is isomorphic with P and no such summand of R is isomorphic with P. Therefore, soc(Q) is a direct sum of simple modules each isomorphic with soc(P) and so Q has no submodule isomorphic with S. Thus, kG/R has no such submodule, so if we set I to be the sum of all submodules of kG isomorphic with S then $I \subseteq R$. Of course, $I \neq 0$: there is an indecomposable projective corresponding to S^* and its dual has socle isomorphic with S; R must have this projective module as a summand. Moreover, I is an ideal of kG. It is already a submodule, that is, left ideal, and whenever ψ is an endomorphism of kG then $\psi(I)$ is also a sum of submodules isomorphic with S, being a homomorphic image of I, so $\psi(I) \subseteq I$. Thus, I is also a right ideal and is an ideal. Let J consist of the endomorphisms of kG whose images lie in I; it is easy to verify that J is an ideal of End(kG).

Let π be the projection of kG onto Q with kernel R. We claim that if $\varphi \in J$ then $\varphi = \varphi\pi - \pi\varphi$. First, $\pi\varphi = 0$ since $\varphi(kG) \subseteq I \subseteq R$ and $\pi(R) = 0$. Second, we assert that $\varphi = \varphi\pi$. Certainly, φ and $\varphi\pi$ coincide on Q as π is the identity on Q. On the other hand, $\pi(R) = 0$ so $\varphi\pi$ vanishes on R. Hence, we need only show that φ also vanishes on R. However, I is a direct sum of modules each isomorphic with S, inasmuch as I is semisimple being the sum of such submodules. Hence if $\varphi(R) \neq 0$ then R has a homomorphic image isomorphic with S. Thus, there is an indecomposable summand of R with such an image; but this must then be an indecomposable projective module isomorphic with P, which contradicts the structure of R.

Let $\varphi \in J$ with $\varphi \neq 0$; this exists as we have seen that kG has a submodule isomorphic with S. Let $\alpha \in$ End(kG) so $\varphi\alpha \in J$ and therefore

$$\varphi\alpha = \varphi\alpha\pi - \pi\varphi\alpha.$$

But endomorphisms are just right multiplications: $\alpha = \rho_a$, $\varphi = \rho_b$, $\pi = \rho_c$, for $a, b, c \in kG$. Thus,

$$ab = cab - abc$$

and

$$
\begin{aligned}
(a, b) &= (ab, 1) \\
&= (cab, 1) - (abc, 1) \\
&= (c, ab) - (ab, c) \\
&= 0.
\end{aligned}
$$

But this holds for all $\alpha \in$ End(kG), that is, for all $a \in kG$. Since the form is non-degenerate, $b = 0$ and $\varphi = 0$, a contradiction.

Exercises

1 Using the description of all indecomposable modules for cyclic groups, determine the dual of each such module.

2 Let Q be an indecomposable injective kG-module with soc$(Q) \cong S$. If U is any kG-module then prove that the multiplicity of S as a composition factor of U equals the dimension of $\mathrm{Hom}_{kG}(U, Q)$.

3 If S and T are simple kG-modules with Q and R corresponding indecomposable projective modules then the multiplicity of S as a composition factor of R equals the multiplicity of T as a composition factor of Q.

7 Tensor products

We introduced tensor products of kG-modules in the last section and we shall now see how useful they are in constructing projective modules. We shall prove some general results and then apply them to the group $\mathrm{SL}(2, p)$ to completely describe its projective modules.

Our first step relates to faithful kG-modules. Here, a kG-module V is *faithful* if each non-identity element of G induces a non-identity linear transformation on V.

Theorem 1 *If V is a faithful kG-module and P is an indecomposable projective kG-module then P is isomorphic with a direct summand of the n-fold tensor product $V \otimes \cdots \otimes V$, for some positive integer n.*

The idea of this section is to use this result by constructing faithful kG-modules whose tensor powers can be readily examined. The proof of the theorem is an application of Galois theory to group theory which should correct the impression that applications only flow in the reverse direction!

Proof Let $R = k[V]$ be the polynomial algebra over k generated by V, so, if $x_1, \ldots, x_r$ is a basis of V then R consists of the polynomials in $x_1, \ldots, x_r$. Since elements of G induce linear transformations of V we now have that they induce automorphisms of R and, in particular, R is a kG-module (but is infinite dimensional). We assert that it suffices to prove that R contains a free submodule. Indeed, suppose this is the case. Since P is isomorphic with a direct summand of any free module, it follows, by the Krull–Schmidt theorem, that P is isomorphic with a direct summand of the kG-module W, which consists of the homogeneous polynomials of degree n, for a positive integer n, in $x_1, \ldots, x_r$. But W is a homomorphic image of the n-fold tensor product $V \otimes \cdots \otimes V$: the linear map which sends each tensor $v_1 \otimes \cdots \otimes v_n$ to $v_1 v_2 \cdots v_n$ in W is readily seen to be a kG-homomorphism. Thus, P is isomorphic with a quotient module of a submodule of $V \otimes \cdots \otimes V$.

However, P is both projective and injective so we have the desired conclusion.

We shall now produce the required free module. Let $K = k(V)$ be the quotient field of R so each non-identity element of G induces a non-identity automorphism of K, as V is faithful. Hence, if F is the fixed field corresponding to G, then K is a Galois extension of F with Galois group the group of automorphisms induced by G. Hence, by the normal basis theorem, there is an element of K whose images under G are not only distinct but form a basis of K as a vector space over F. If f_1/f_2 is such an element, where $f_1, f_1 \in R$, then multiplying this element by $\prod_{g \in G} g(f_2)$, which is in F, we get a polynomial f which has the same property. Since $F \supseteq k$, we certainly have that the images $g(f)$ of f under the elements of G are linearly independent over k. But these images clearly form a kG-module isomorphic with the kG-module kG and the proof is complete.

Proposition 2 *If G has no non-identity normal p-subgroups and $S_1, \ldots, S_n$ are the simple kG-module then $S_1 \oplus \cdots \oplus S_n$ is a faithful kG-module.*

Combining this result, the previous one and the Krull–Schmidt theorem, we have that if P is an indecomposable projective kG-module then P is isomorphic with a summand of a tensor product $S_{i_1} \otimes \cdots \otimes S_{i_t}$ so we need only analyze the structure of tensor products of simple modules to get at the projective modules for most groups.

Proof The set of elements of G which induce the identity linear transformation on each simple kG-module forms a normal subgroup; hence, if $g \in G$ is such an element then we need only see that g is a p-element, that is, has order a power of p. However, $g - 1$ now annihilates each simple kG-module so $g - 1 \in \mathrm{rad}(kG)$. Hence, $(g - 1)^r = 0$, for some positive integer r, so $(g - 1)^{p^r} = 0$, that is, $g^{p^r} = 1$.

In analyzing tensor products, the most useful tool is the next result.

Lemma 3 *If U, V and W are kG-modules then*
$$\mathrm{Hom}_{kG}(U \otimes V, W) \cong \mathrm{Hom}_{kG}(U, V^* \otimes W).$$

In order to establish this result, we shall first show how the functor Hom can be used to construct modules. If U and V are kG-modules then $\mathrm{Hom}_k(U, V)$, the vector space of linear transformations from U to V, is also, in a natural way, a kG-module. Indeed, if $\varphi \in \mathrm{Hom}_k(U, V)$ and $g \in G$ then $g\varphi$ is defined by setting $(g\varphi) = g(\varphi(g^{-1}u))$ whenever $u \in U$ and it is trivial to verify that this defines a module. Note that if $V = k$, the trivial module, then $(g\varphi)u = \varphi(g^{-1}(u))$ and $\text{Hom}_k(U, k) \cong U^*$ as kG-modules. Now $\text{Hom}_{kG}(U, V)$ is a subspace of $\text{Hom}_k(U, V)$. Indeed, if $\varphi \in \text{Hom}_k(U, V)$ then φ commutes with the action of $g \in G$ if, and only if, $g\varphi = \varphi$, since φ and g commute exactly when $\varphi(u) = g(\varphi(g^{-1}(u)))$ for all $u \in U$. Thus, $\text{Hom}_{kG}(U, V)$ consists of the elements of the kG-module $\text{Hom}_k(U, V)$ which are left fixed by G.

This new construction is really just an old one presented in a useful way. In fact, $\text{Hom}_k(U, V) \cong U^* \otimes V$, as kG-modules. Recall that there is a natural isomorphism of the vector space $U^* \otimes V$ onto the vector space $\text{Hom}_K(U, V)$ which maps $\varphi \otimes v$, $\varphi \in U^*$, $v \in V$, to the linear transformation which sends $u \in U$ to $\varphi(u)v$ in V. Again, it is easy to demonstrate that this is a kG-isomorphism. Indeed, the image of $\varphi \otimes v$, when g is applied to it, is the linear transformation sending u to $g(\varphi(g^{-1}u)v)$, which is $\varphi(g^{-1}u)gv$, while $g\varphi \otimes gv$ is sent to the linear transformation sending u to $(g\varphi)(u)gv = \varphi(g^{-1}u)gv$.

Now we are ready to prove Lemma 3. It suffices to establish a stronger result: the modules $\text{Hom}_k(U \otimes V, W)$ and $\text{Hom}_k(U, V^* \otimes W)$ are isomorphic. An isomorphism will then map the fixed points of G in the one space to the fixed points in the other space. However,

$$\text{Hom}_k(U \otimes V, W) \cong (U \otimes V)^* \otimes W$$
$$\cong U^* \otimes (V^* \otimes W)$$
$$\cong \text{Hom}_k(U, V^* \otimes W).$$

One other useful fact is as follows.

Lemma 4 *If V is a kG-module and P is a projective kG-module then $V \otimes P$ is also projective.*

Proof If P is a direct summand of the free module F then $V \otimes P$ is a summand of $V \otimes F$; hence it suffices to show that $V \otimes kG$ is a free module. If $0 \neq v \in V$ then $v \otimes 1$ generates a free submodule of $V \otimes kG$: the images of $g(v \otimes 1) = gv \otimes g$, as g runs over G, have linearly independent second factors so are linearly independent. Let $v_1, \ldots, v_n$ be a basis for V and let F_i be the submodule of $V \otimes kG$ generated by $v_i \otimes 1$, $1 \leqslant i \leqslant n$ so $F_i \cong kG$. Since $V \otimes kG$ has dimension $n|G|$ we will have that $V \otimes kG$ is the direct sum of F_1, $\ldots, F_n$ once we prove that $V \otimes kG$ is their sum. However, if $v \in V$ and $g \in G$, choose $\alpha_1, \ldots, \alpha_n \in k$ such that $g^{-1}v = \alpha_1 v_1 + \cdots + \alpha_n v_n$ and then

$$g(\alpha_1(v_1 \otimes 1) + \cdots + \alpha_n(v_n \otimes 1)) = g(\alpha_1 v_1 + \cdots + \alpha_n v_n) \otimes g$$
$$= v \otimes g.$$

It is also easy to get this result as a consequence of the basic results of the next section. Moreover, one can give a proof, without referring to a basis, in the spirit of our approach but at the expense of a longer argument (see the exercises).

For the remainder of the section we set $G = \mathrm{SL}(2, p)$ and let $V_1, \ldots, V_p$ be the simple kG-modules just as before. There is one fact that we need that depends on a general result to be proved in a later section: V_p is projective. We shall establish now that V_p is projective when considered as a module for a Sylow p-subgroup of G; later we shall see that this implies the claimed projectivity. Now G is of order $(p^2 - 1)p$ and the matrix

$$\begin{pmatrix} 1 & 1 \\ 0 & 1 \end{pmatrix}$$

is an element of order p of G so it generates a Sylow p-subgroup of G. However, we have already established that V_p is a uniserial module for this subgroup of order p. Since V_p is p-dimensional, it is the indecomposable projective module for this cyclic group.

Let $P_1, \ldots, P_p$ be the indecomposable projective modules corresponding to $V_1, \ldots, V_p$, respectively, so that $P_p \cong V_p$. We shall establish the structure of each of these projective modules and shall achieve the following results. First, P_1 is uniserial, with three composition factors which occur as V_1, V_{p-2}, V_1 (unless $p = 2$ in which case there are just two factors V_1 and V_1) in the unique composition series. Hence, we can picture P_1 as follows:

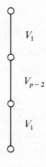

If $p > 2$ then P_{p-1} is also uniserial of length three and can be pictured as

follows

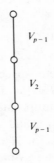

Finally, if $1 < n < p-1$ then $\mathrm{rad}(P_n)/\mathrm{soc}(P_n) \cong V_{p+1-n} \oplus V_{p-1-n}$. Now $P_n/\mathrm{rad}(P_n) \cong \mathrm{soc}(P_n) \cong V_n$ so we have P_n as pictured in three layers, V_n, $V_{p+1-n} \oplus V_{p-1-n}$, V_n. Since V_{p+1-n} and V_{p-1-n} are distinct simple modules their direct sum has exactly four submodules: zero, each of them, their sum. Since $\mathrm{rad}(P_n)$ is the unique maximal submodule of P_n and $\mathrm{soc}(P_n)$ is the unique minimal submodule of P_n, the full lattice of submodules of P_n is as follows, where the labels give the isomorphism class of quotients of successive factors:

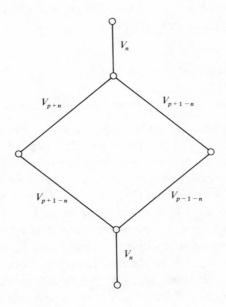

In order to establish these structures we must first analyze the structure of certain tensor products of simple modules.

Lemma 5 *If* $2 \leqslant n < p$ *then* $V_2 \otimes V_n \cong V_{n-1} \oplus V_{n+1}$.

Proof Remember that V_n is the vector space of homogeneous polynomials in X and Y of degree $n - 1$. Since G acts as automorphisms of $k[X, Y]$, the map of $V_2 \otimes V_n$ to V_{n+1} given by multiplication of polynomials (so $f \otimes g$ is sent to fg) is a module homomorphism and its image is all of V_{n+1} since any monomial in V_{n+1} has X or Y as a factor. Hence, in order to show the kernel of this multiplication homomorphism is isomorphic with V_{n-1}, it is enough to define a one-to-one homomorphism of V_{n-1} to $V_2 \otimes V_n$ which has image in the kernel, since the kernel has dimension $2n - (n+1) = n - 1$. Define a linear transformation φ of V_{n-1} to $V_2 \otimes V_n$ by $\varphi(f) = X \otimes Yf - Y \otimes Xf$ so the image of φ is in the kernel. Moreover, φ is a module homomorphism. If

$$s = \begin{pmatrix} a & b \\ c & d \end{pmatrix}$$

is an element of G then $ad - bd = 1$ and

$$
\begin{aligned}
s(\varphi(f)) &= s(X \otimes Yf - Y \otimes Xf) \\
&= s(X) \otimes s(Y)s(f) - s(Y) \otimes s(X)s(f) \\
&= (aX + cY) \otimes (bX + dY)s(f) - (bX + dY) \otimes (aX + cY)s(f) \\
&= (ad - bc)(X \otimes Ys(f) - Y \otimes Xs(f)) \\
&= \varphi(s(f)).
\end{aligned}
$$

Furthermore, φ is one-to-one. The elements $X^i Y^{n-2-i}, 0 \leqslant i \leqslant n-2$, form a basis of V_{n-1} and

$$\varphi(X^i Y^{n-2-i}) = X \otimes X^i Y^{n-1-i} - Y \otimes X^{i+1} Y^{n-2-i}$$

and the $2(n-1)$ tensors, whose differences are the $n-1$ images, are themselves a basis of $V_2 \otimes V_n$ and so linearly independent.

To prove the lemma, it is now only necessary to prove that $V_2 \otimes V_n$ has a submodule isomorphic with V_{n+1} as then it has submodules isomorphic with each of the simple modules V_{n-1} and V_{n+1} and their dimensions add up to the dimension of $V_2 \otimes V_n$. We shall prove this by downward induction on n. If $n = p - 1$ then $V_2 \otimes V_n$ has the projective module V_p as a homomorphic image so it must have a submodule isomorphic with V_p. Now suppose that $n + 1 < p$ and $V_2 \otimes V_{n+1}$ is as claimed. It suffices, since V_{n+1} is a simple module, to prove that $\mathrm{Hom}_{kG}(V_{n+1}, V_2 \otimes V_n) \neq 0$. However, the dual V_2^* is also two-dimensional and simple and so must be isomorphic with V_2, the only such module, and we have

$$\mathrm{Hom}_{kG}(V_{n+1}, V_2 \otimes V_n) \cong \mathrm{Hom}_{kG}(V_{n+1} \otimes V_2^*, V_n)$$
$$\cong \mathrm{Hom}_{kG}(V_{n+1} \otimes V_2, V_n)$$
$$\cong \mathrm{Hom}_{kG}(V_n \oplus V_{n+2}, V_n)$$
$$\neq 0.$$

We are now prepared to establish the structure of the projective modules. The module $V_2 \otimes V_p$ is projective, by Lemma 4 as V_p is projective, and the first part of the proof of Lemma 5 applies: $V_2 \otimes V_p$ has a submodule isomorphic with V_{p-1} and the quotient of this submodule is isomorphic with V_{p+1} (which is defined and not one of the simple modules). Therefore, $V_2 \otimes V_p$ has a summand isomorphic with P_{p-1}: for $V_2 \otimes V_p$ has a summand which is indecomposable and has V_{p-1} isomorphic with a submodule (and hence all) of its socle. Now $P_{p-1} \not\supseteq \mathrm{soc}(P_{p-1})$, since V_{p-1} cannot be projective as its dimension is not divisible by p, so $P_{p-1} \supseteq \mathrm{rad}(P_{p-1}) \supseteq \mathrm{soc}(P_{p-1})$ and P_{p-1} has dimension at least $2(p-1)$. Since p divides the dimension of P_{p-1} and $V_2 \otimes V_p$ is $2p$-dimensional, we have $V_2 \otimes V_p \cong P_{p-1}$, if $p > 2$. (If $p = 2$ it is easy to describe $P_{p-1} = P_1 : 6 = |G| = \dim P_1 + 2 \dim P_2$ so P_1 is two-dimensional and its structure is immediate.) In this case, the submodule $V = \mathrm{soc}(V_2 \otimes V_p) \cong V_{p-1}$ and $V_2 \otimes V_p / U \cong V_{p-1}$ where $U = \mathrm{rad}(V_2 \otimes V_p)$. Hence, we have the following picture.

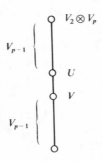

Hence, U/V is two-dimensional. Since U is the unique maximal submodule of $V_2 \otimes V_p$ and V is the unique minimal submodule we need only show that $U/V \cong V_2$ to prove that P_{p-1} is uniserial as described. But $V_2 \otimes V_p / V \cong V_{p+1}$ and U/V is the unique maximal submodule of this quotient so we need only demonstrate that V_{p+1} has a submodule isomorphic with V_2. The taking of pth powers is an isomorphism of $k[X, Y]$ into itself so it is a kG-module homomorphism and one-to-one. Since it maps V_2 into V_{p+1}, as X and Y are sent to X^p and Y^p, respectively, we are done.

We proceed to study P_{p-2} and we do this by examining $V_2 \otimes P_{p-1}$. This module is projective and has a series of submodules whose successive quotients are isomorphic with $V_2 \otimes V_{p-1}$, $V_2 \otimes V_2$ and $V_2 \otimes V_{p-1}$, that is, $V_{p-2} \oplus V_p$, $V_1 \oplus V_3$, $V_{p-2} \oplus V_p$. Since V_p is projective and since V_{p-2} is a homomorphic image, we have that $V_2 \otimes P_{p-1}$ has a direct summand isomorphic with $P_{p-2} \oplus V_p \oplus V_p$ (and with $P_1 \oplus V_3 \oplus V_3 \oplus V_3$ if $p=3$). But P_{p-2} has two composition factors isomorphic with V_{p-2}, its dimension is divisible by p and its only other possible composition factors are V_1 and V_3 (just V_1 if $p=3$) so we deduce that $\text{rad}(P_{p-2})/\text{soc}(P_{p-2})$ has V_1, V_3 as composition factors (and just V_1 if $p=3$ so there $P_{p-2} = P_1$ is as claimed). We need just see that this quotient is isomorphic with $V_1 \oplus V_3$. If it is not semisimple then it must be uniserial, since it has just two composition factors, and so it will not be isomorphic with its dual, which would be uniserial but with composition factors in reverse order. However, P_{p-2} is isomorphic with its dual (as P_{p-2}^* is the indecomposable projective corresponding with $\text{soc}(P_{p-2})^* \cong V_{p-2}^* \cong V_{p-2}$). The relations between radicals, socles and duality immediately imply that $\text{rad}(P_{p-2})/\text{soc}(P_{p-2})$ is isomorphic with its dual, and we have established the structure of P_{p-2}.

Next, suppose that P_n, $2 < n < p-1$ is just as claimed. Therefore $V_2 \otimes P_n$ has a series of submodules with successive quotients $V_2 \otimes V_n$, $V_2 \otimes (V_{p+1-n} \oplus V_{p-1-n})$, $V_2 \otimes V_n$, that is, $V_{n-1} \oplus V_{n+1}$, $V_{p+2-n} \oplus V_{p-n} \oplus V_{p-n} \oplus V_{p-n-2}$, $V_{n-1} \oplus V_{n+1}$, where in the case of $n = p-2$ the term V_{p-n-2} should be deleted. Since we have already dealt with P_{p-1}, P_{p-2} we can proceed by downward induction, so P_{n+1} is also to be assumed as claimed. Since $V_2 \otimes P_n$ is projective and has $V_{n+1} \oplus V_{n-1}$ as a homomorphic image we have that $V_2 \otimes P_n \cong P_{n+1} \oplus U$, where U is projective, has V_{n-1} as a homomorphic image and has four composition factors, namely $V_{n-1}, V_{p+2-n}, V_{p-n}, V_{n-1}$. Proceeding now exactly as in the previous paragraph we obtain that $U \cong P_{n-1}$ and P_{n-1} is as claimed.

Finally, it remains to deal with P_1 by examining $V_2 \otimes P_2$. Here again we get a series of submodules but with V_p a summand of the middle factor and so using the projectivity of V_p and the preceding argument we get

$$V_2 \otimes P_2 \cong P_3 \oplus V_p + U$$

where U is projective, has V_1 as a homomorphic image, has V_1, V_{p-2}, V_1 as composition factors and again we have the desired structure and our argument is totally complete.

Exercises

1   If U is a kG-module and P is a projective kG-module with $U/\text{rad}(U) \cong P/\text{rad }P$ then U is a homomorphic image of P.

2   If S is a simple kG-module then the multiplicity of S as a composition factor of $U/\mathrm{rad}(U)$, for a kG-module U, equals the dimension of $\mathrm{Hom}_{kG}(U, S)$.

3   The multiplicity of S, a simple kG-module, as a composition factor of the quotient of $V \otimes kG$ by its radical, where V is a kG-module, is $\dim V \cdot \dim S$. Deduce that $V \otimes kG \cong kG \oplus \cdots \oplus kG$ (dim V times).

4   If p does not divide the dimension of the kG-module V then $V \otimes V^*$ has a summand isomorphic with the trivial module. (Hint: $V \otimes V^* \cong \mathrm{Hom}_k(V, V)$; consider scalar multiples of the identity linear transformation and linear transformations of trace zero.)

5   Now, and for the remaining exercises, assume that $G = \mathrm{SL}(2, p)$. If U is a kG-module with $U \cong U^*$ and $\mathrm{End}(U) \cong k$ then U is simple.

6   If the indecomposable kG-module U is a direct summand of a tensor product of simple kG-modules then U is a summand of $V_2 \otimes \cdots \otimes V_2$ for some number of factors.

7   Show there are at most $2p - 1$ modules in the previous exercise.

8   Show that P_1 is a summand of $V_2 \otimes \cdots \otimes V_2$, $2p - 1$ factors, and that it is not a summand of a tensor product of a smaller number of such factors.

III

Modules and subgroups

Up to this point subgroups have played no substantial part but the deepest theorems bring in p-local subgroups in an essential way. In fact, the connections between kG-modules and modules for local subgroups are the most important tools. This chapter is devoted to introducing and developing these ideas.

8 Induced modules

The notion of a free module depended on a subspace of a module. However, a subspace is nothing more than a submodule for the identity subgroup and we shall generalize the idea of a free module by using modules for subgroups in place of subspaces. This easy extension of the idea of a free module is the basic means of relating kG-modules and modules for subgroups. After introducing this notion in a formal way, we shall give the usual description in terms of tensor products.

We now say that the kG-module is *relatively H-free*, where H is a subgroup of G, if there is a kH-submodule X of U such that any kH-homomorphism of X to any kG-module V extends uniquely to a kG-homomorphism of U to V. Notice that, as we promised, if $H = 1$ then this notion is the idea of a free module. We also say that U is relatively H-free with respect to X and, with abuse of notation, with respect to any X', a kH-module isomorphic to X. The first result is entirely analogous to the one on free modules and the proof can be instantly supplied by mimicking the early one.

Lemma 1 *If U and V are relatively H-free kG-modules with respect to the kH-submodules X and Y, respectively, and X is isomorphic with Y, then U is isomorphic with V.*

Proposition 2 *If X is a kH-module, where H is a subgroup of G, then there is a kG-module which is relatively H-free with respect to X.*

To understand the idea of the proof, let us examine the structure of the kG-module U which is relatively H-free with respect to X. Set $U_1 = \sum_{s \in G/H} sX$ so U_1 is certainly a kG-submodule since X is a kH-submodule. The natural map of U onto U/U_1 and the zero map of U to U/U_1 are both extensions of the zero homomorphism of X to U/U_1 so, by the uniqueness part of the defining property of U, we have $U/U_1 = 0$, that is, $U_1 = U$. We must simply construct a module like U_1 and actually make the sum defining U_1 a direct one. This is not unexpected, since in the case that $H = 1$ and X is one-dimensional we have $U \cong kG$ which is a direct sum of one-dimensional subspaces, one for each element of G.

Hence, let S be a set of (left) coset representatives of H in G, and, for convenience, let $1 \in S$. For each $s \in S$ the cartesian product $\{s\} \times X$ can be given the structure of a vector space by using the operations on X. Form the direct sum of these subspaces and, with an abuse of notation, consider this as all sums $\sum_{s \in S} (s, x_s)$. We shall now give this a kG-module structure and show it has the required properties. If $g \in G$, $s \in S$, $x \in X$ then set $g(s, x) = (t, hx)$, where $gs = th$, $t \in S$, $h \in H$ and let g act additively on sums of such elements. It is easy to see this defines a module. In fact, if $g' \in G$ and $g't = uh'$, where $u \in S$, $h' \in H$ then

$$g'(g(s, x)) = g'(t, hx)$$
$$= (u, h'hx)$$
$$= (g'g)(s, x)$$

since $g'gs = g'th = uh'h$. Suppose that φ is a kH-homomorphism of $\{1\} \times X$ to the kG-module V; note that $\{1\} \times X$ is a kH-submodule and naturally isomorphic with X. Define a linear transformation Φ by demanding that $\Phi((s, x)) = s\varphi(x)$. This extends φ and it is also a kG-homomorphism:

$$\Phi(g(s, x)) = \Phi((t, hx))$$
$$= t\varphi(hx)$$
$$= th\varphi(x)$$
$$= g(s\varphi(x))$$
$$= g\Phi((s, x)).$$

Moreover, if Ψ is any such extension of φ then $\Psi((s, x)) = \Psi(s(1, x)) = s\Psi((1, x)) = s\varphi(x) = \Phi((s, x))$, so the proposition is established.

We can now give a very useful characterization of relatively free modules.

Corollary 3 *If U is a kG-module generated by the kH-submodule X, for a subgroup H of G, then U is relatively H-free with respect to X if, and only if,*

$$\dim U = |G:H| \dim X.$$

Proof If U is relatively H-free with respect to X then it is isomorphic with the module just constructed and so is the direct sum of $|G:H|$ subspaces each of which has the same dimension as X and $\dim U$ is as stated. On the other hand, suppose that U has dimension $|G:H| \dim X$. Let U' be a relatively H-free kG-module with respect to the kH-module X' with $X' \cong X$. Our assumption implies that $\dim U = \dim U'$. Hence, we need only show there is a kG-homomorphism of U onto U' extending an isomorphism of X onto X'. But any kG-homomorphism of U to U' extending an isomorphism of X' onto X has image in U a kG-submodule containing X and so must be all of U, by assumption.

We now turn to the usual tensor product description of these relatively free modules. Let V be a kH-module, where H is our usual subgroup of G, and form the tensor product $kG \otimes V$ of two vector spaces (which we now just consider as a vector space) and let $kG \otimes_{kH} V$ be the quotient by the subspace spanned by all elements of the form $ah \otimes v - a \otimes hv$, where $a \in kG$, $v \in V$, $h \in H$. We shall now give this quotient space the structure of a kG-module. If $a \in kG$, $v \in V$ and $g \in G$ define $g(a \otimes v) = ga \otimes v$. The bilinearity of this action means we have defined a kG-module structure on the vector space $kG \otimes V$ (and it is not the usual structure) and since

$$g(ah \otimes v - a \otimes hv) = (ga)h \otimes v - (ga) \otimes hv$$

the subspace given is preserved by the action of g so we have a kG-module structure on $kG \otimes_{kH} V$. This module is denoted by V^G and it is called the kG-module *induced* by the kH-module V. If $a \in kG$ and $v \in V$ then we also abuse the notation and write $a \otimes v$ for the element of V^G which corresponds to this tensor. With these definitions we have the first important remark.

Lemma 4 *The induced module V^G is relatively H-free with respect to V. Moreover, V^G is the vector space direct sum*

$$V^G = \sum_{s \in G/H} s \otimes V$$

and each subspace $s \otimes v$ has dimension equal to that of V.

Proof Let U be a relatively H-free kG-module with respect to V so by

Lemma 1 and the construction of Proposition 2 we have that

$$U = \sum_{s \in G/H} sV.$$

Define a linear transformation φ of the vector space $kG \otimes V$ to U such that $\varphi(g \otimes u) = gv$, if $g \in G$, $v \in V$. The bilinearity of this equation guarantees the existence of such a map. The expression of U as a sum just given shows that the image of φ is all of U. If $h \in H$ then

$$\varphi(gh \otimes v - g \otimes hv) = ghv - ghv = 0$$

so φ defines a linear transformation Φ of the quotient V^G onto U. Moreover, if $g' \in G$ then

$$\Phi(g'(g \otimes v)) = \Phi(g'g \otimes v)$$
$$= g'gv$$
$$= g'\Phi(g \otimes v)$$

so Φ is a kG-homomorphism of V^G onto U. Since in V^G we have $gh \otimes v = g \otimes hv$ and $hv \in V$, it follows that $V^G = \sum_{s \in G/H} s \otimes V$ and so dim $V^G \leqslant |G : H|$ dim $V =$ dim U. Thus, Φ must be an isomorphism of V^G onto U. Since

$$\Phi(s \otimes v) = sV$$

all the statements are fully proved as V^G now is relatively kH-free with respect to $1 \otimes V$, the sum expressing V^G is direct and each $s \otimes v$ has dimension that of sV.

Our next task is to prove an Omnibus Lemma which is a collection of basic facts about induction. We have divided this result into two parts. The first does not deal with maps while the second does.

Lemma 5 *Let V, V_1, V_2 be kH-modules for the subgroup H of G and let U be a kG-module.*

(1) *If V is free (projective) then V^G is free (projective).*
(2) $(V_1 \oplus V_2)^G \cong V_1^G \oplus V_2^G$
(3) $(V^*)^G \cong (V^G)^*$.
(4) *If W is a kL-module for a subgroup L of H, then $(W^H)^G \cong W^G$.*
(5) $U \otimes V^G \cong (U_H \otimes V)^G$.

Proof The direct sum $V_1^G \oplus V_2^G$ contains $V_1 \oplus V_2$ as a kH-submodule and is generated by it. Corollary 3 now gives us that (2) holds. As a consequence, in order to prove (1), it is enough to do the case of free modules. However, if V is a free kH-module, that is, a kH-module which is relatively 1-free, then $V \cong X^H$ where X is a k-module. Thus, $V^G \cong (X^H)^G$ and X^G is a free kG-module, so we need only establish (4) to show that (1) holds.

Let U be relatively H-free with respect to V. We shall show that U^* is relatively H-free with respect to V^*. This implies that (3) holds: since $U \cong V^G$ we still have $(V^G)^* \cong U^* \cong (V^*)^G$. Now $U = \sum sV$ is a direct sum, where s runs over a set S of (left) coset representatives for H in G. Let $\hat{V} \subseteq U^*$ be the subspace of linear functionals which vanish on sV if $s \notin H$ so clearly $\hat{V}$ is a kH-submodule of U^* and $\hat{V} \cong V^*$. Since dim $U = \dim U^*$ and dim $V = \dim \hat{V}$, we have, by Corollary 3, that it is sufficient to prove that U^* is generated as a kG-module by $\hat{V}$. If $s, t \in S$, $v \in V$ and $\varphi \in \hat{V}$ then $(t\varphi)(sv) = \varphi(t^{-1}sv)$ so this will be zero if $tH \neq sH$. Thus $t\hat{V}$ consists of the linear functionals which vanish on all sV for $sH \neq tH$. However, since $U = \sum sV$ is a direct sum, every linear functional is a sum of such elements and so (3) holds.

We turn to (4) now. First, note that

$$\dim (W^H)^G = |G:H| \dim W^H$$

$$= |G:H||H:L| \dim W$$

$$= |G:L| \dim W$$

so, by Corollary 3, we need only show that $(W^H)^G$ is generated as a kG-module by a kL-module isomorphic with W. But $1 \otimes W \cong W$ generates $(W^H)^G$ as kG-module so (4) is established.

In (5) the tensor product $U \otimes V^G$ is the usual one for kG-modules so $U \otimes V^G$ contains $U_H \otimes V$ as a kH-submodule (after identifying V and $1 \otimes V \subseteq V^G$) and

$$\dim (U \otimes V^G) = (\dim U)(\dim V^G)$$

$$= |G:H| \dim (U_H \otimes V)$$

so, by Corollary 3 we just need to show that $U_H \otimes V$ generates $U \otimes V^G$. But, if $u \in U$, $g \in G$ and $v \in V$ then

$$u \otimes gv = g(g^{-1}u \otimes v) \in g(U \otimes V)$$

so all parts of the lemma are proved.

We now turn to the second part of the Omnibus Lemma.

Lemma 6 *With the above notation, we also have the following:*

(1) $\operatorname{Hom}_{kG}(V^G, U) \cong \operatorname{Hom}_{kH}(V, U_H)$.

(2) $\operatorname{Hom}_{kG}(U, V^G) \cong \operatorname{Hom}_{kH}(U_H, V)$.

(3) *If* $\gamma \in \operatorname{Hom}_{kH}(U_H, V)$ *then the map sending* $u \in U$ *to* $\sum_{s \in G/H} s \otimes \gamma(s^{-1}u)$ *is in* $\operatorname{Hom}_{kG}(U, V^G)$ *and this yields an isomorphism of* $\operatorname{Hom}_{kH}(U_H, V)$ *onto* $\operatorname{Hom}_{kG}(U, V^G)$.

(4) *If* $\alpha \in \mathrm{Hom}_{kH}(V_1, V_2)$ *then there is a unique* $\alpha^G \in \mathrm{Hom}_{kG}(V_1^G, V_2^G)$ *which extends* α.

(5) *If the short exact sequence*

$$0 \to V_1 \overset{\alpha}{\to} V \overset{\beta}{\to} V_2 \to 0$$

is exact then so is

$$0 \longrightarrow V_1^G \overset{\alpha^G}{\longrightarrow} V^G \overset{\beta^G}{\longrightarrow} V_2^G \longrightarrow 0$$

and one is split if, and only if, the other is split.

Recall that to say the first sequence of modules and module homomorphisms is exact means that α is one-to-one, $\alpha(V_1)$ is the kernel of β and $\beta(V) = V_2$ so α and β show that V has a submodule isomorphic with V_1 with corresponding quotient isomorphic with V_2. As before, this splits if the map β splits (or equivalently if $\alpha(V_1)$ is a direct summand of V).

Proof The first assertion is just a consequence of the mapping property of relatively free modules: if $\alpha \in \mathrm{Hom}_{kH}(V, U_H)$ then α has a unique extension $\hat{\alpha} \in \mathrm{Hom}_{kG}(V^G, U)$ and any element of $\mathrm{Hom}_{kG}(V^G, U)$ so arises because it is the extension of its restriction to V. Using the previous lemma we now easily obtain (2):

$$\mathrm{Hom}_{kG}(U, V^G) \cong \mathrm{Hom}_{kG}((V^G)^*, U^*)$$
$$\cong \mathrm{Hom}_{kG}((V^*)^G, U^*)$$
$$\cong \mathrm{Hom}_{kH}(V^*, (U^*)_H)$$
$$\cong \mathrm{Hom}_{kH}(U_H, V).$$

We turn to (3) where to each $\gamma \in \mathrm{Hom}_{kH}(U_H, V)$ we have defined a linear transformation γ' from U to V^G. We must prove that γ' is a kG-module homomorphism and that the map which sends γ to γ' is a vector space isomorphism of $\mathrm{Hom}_{kH}(U_H, V)$ onto $\mathrm{Hom}_{kG}(U, V^G)$, so making statement (2) explicit. However, the formula for γ' shows that the map associating γ' to γ is linear and it is one-to-one as $\gamma \neq 0$ forces $\gamma' \neq 0$ since $\gamma'(u) = 1 \otimes \gamma(u) + \cdots$ has first term $\gamma(u)$. But $\mathrm{Hom}_{kG}(U, V^G)$ and $\mathrm{Hom}_{kH}(U, V)$ have the same dimension, by (2), so the map sending γ to γ' will have image all of $\mathrm{Hom}_{kG}(U, V^G)$ once we show that γ' is, in fact, a module homomorphism.

However, if $g \in G$ and $u \in U$ then

$$\gamma'(gu) = \sum_{s \in G/H} s \otimes \gamma(s^{-1}gu)$$

$$= \sum g(g^{-1}s) \otimes \gamma(s^{-1}gu)$$

$$= g \sum g^{-1}s \otimes \gamma(s^{-1}gu)$$

$$= g \sum_{t \in G/H} t \otimes \gamma(t^{-1}u)$$

because the elements $g^{-1}s$ give another set of coset representatives. However, if $t = sh_t$, $h_t \in H$, then

$$\sum t \otimes \gamma(t^{-1}u) = \sum sh_t \otimes \gamma(h_t^{-1}su)$$

$$= \sum s \otimes h_t h_t^{-1} \gamma(su)$$

$$= \sum s \otimes \gamma(su)$$

$$= \gamma'(u)$$

so $\gamma'(gu) = g\gamma'(u)$ as claimed.

The fourth part of the lemma is just a special case of the defining property of relatively free modules. If $\alpha \in \mathrm{Hom}_{kH}(V_1, V_2)$ then α can be regarded as having image $1 \otimes V_2 \subseteq V_2^G$, a kG-module, so there is a unique extension of a kG-module homomorphism of V_1^G to V_2^G. It is also clear that for $v_s \in V$

$$\alpha^G \left(\sum_{s \in G/H} s \otimes v_s \right) = \sum_{s \in G/H} s \otimes \alpha(v_s),$$

as $\alpha^G(s \otimes v_s) = \alpha^G(s(1 \otimes v_s)) = s\alpha^G(1 \otimes v_s) = s(1 \otimes \alpha(v_s))$.

This leaves us the task of demonstrating the truth of part (5). The exactness of the second sequence is easy to see. Indeed, by the formula just given for α^G, which applies equally to β^G, we have $\alpha^G(s \otimes V) = s \otimes \alpha(V_1)$ and the kernel of β^G on $s \otimes V$ is just $x \otimes \alpha(V_1)$ as $\alpha(V_1)$ is the kernel of β. The one-to-one property of α^G and the surjectivity of β^G are also clear. Moreover, suppose the first sequence splits: let $\gamma \in \mathrm{Hom}_{kH}(V_2, V)$ with $\beta\gamma$ the identity of V_2. Thus, $\beta^G\gamma^G$ extends $\beta\gamma$ to a kG-module homomorphism of V_2^G to V_2^G and so must be the identity there by the uniqueness part of (4).

Finally, suppose that $\psi \in \mathrm{Hom}_{kG}(V_2^G, V^G)$ and $\beta^G\gamma$ is the identity of V_2. Let π be the kH-module homomorphism of V^G onto V which satisfies $\pi(1 \otimes v) = v$ and $\pi(s \otimes v) = 0$ if $s \notin H$. Define a map φ from V_2 to V by

$$\varphi(v_2) = \pi\psi(1 \otimes v_2)$$

so φ is a kH-module homomorphism. We assert that $\beta\varphi$ is the identity of V_2. Let $v_2 \in V_2$ so

$$\psi(1 \otimes v_2) = \sum s \otimes u_s$$

where s sums over a set of coset representatives, one of which is 1, and $u_s \in V$.
Thus

$$\beta^G(\psi(1 \otimes v_2)) = \sum s \otimes \beta(u_s)$$

and this must be just $1 \otimes v_2$ so $\beta(u_1) = v_2$. Hence

$$\varphi(v_2) = \pi\psi(1 \otimes v_2)$$

$$= \pi\left(\sum s \otimes u_s\right)$$

$$= u_1$$

so $\beta(\varphi(v_2)) = V_2$ as claimed.

The next result is Mackey's theorem, which is a fundamental lemma. It allows us to describe the process of inducing and then restricting in terms of restriction followed by induction so that we can relate induction to G to constructions which usually deal with just proper subgroups of G.

For this result we fix two subgroups of G, namely H and L. Recall that if $g \in G$ then the subset LgH is called a double coset and G is the disjoint union of these subsets. Let S be a set of double coset representatives for H, L in G so G is the disjoint union of the subsets LsH as s runs over S (sometimes expressed $G = \bigcup_{s \in L\backslash G/H} LsH$). If V is a kH-module then $s \otimes V$ is a summand of V^G and it is a module for sHs^{-1}. Indeed, if $h \in H$, $v \in V$

$$shs^{-1}(s \otimes v) = sh \otimes v$$

$$= s \otimes hv.$$

Thus, shs^{-1} acts on $s \otimes v$ according as h acts on v. This is an example of transport of structure: conjugation by s carries H to sHs^{-1} and $s \otimes V$ is the module for shs^{-1} corresponding to the kH-module V. We denote by $s(V)$ any $k[sHs^{-1}]$-module corresponding to the kH-module V so $s(V) \cong s \otimes V$. Now $L \cap sHs^{-1}$ is a subgroup of sHs^{-1} so we can restrict $s(V)$ to this subgroup and since it is contained in L we can then induce to L. We can now state Mackey's theorem.

Lemma 7 *With the above notation*

$$(V^G)_L \cong \bigoplus_{s \in L\backslash G/H} (s(V)_{L \cap sHs^{-1}})^L.$$

The proof is just a careful analysis of the connection between cosets and double cosets and a careful keeping track of modules.

Proof For each $s \in S$, let T_s be a set of (left) coset representatives for the subgroup $L \cap sHs^{-1}$ in L. It is an elementary fact from group theory, and easily verified, that LsH is the disjoint union of the cosets tsH as t runs over T_s. Therefore, V^G is the direct sum of all the subspaces $ts \otimes V$ as s runs over S and t over T_s. Since each t is in L, the sum of all $ts \otimes V$, for a fixed $s \in S$ and all $t \in T_s$, is a kL-module, call it V_s, and $(V^G)_L$ is the direct sum of all the V_s, $s \in S$. Hence, it suffices to prove that $V_s \cong (s(V)_{L \cap sHs^{-1}})^L$.

But dim $V_s = |L: L \cap sHs^{-1}|$ dim V since $|T_s| = |L: L \cap sHs^{-1}|$. Moreover, $s \otimes V \subseteq V_s$ and $s \otimes V$ as $k[L \cap sHs^{-1}]$-module is isomorphic with $s(V)$ as a $k[L \cap sHs^{-1}]$-module. Hence, it suffices to show that V_s is generated as a kL-module by $s \otimes V$, in view of Corollary 3. But $LsH = T_s sH$ so this is clear.

The final result of this section is Green's indecomposability criterion which is about induction in the case of normal subgroups.

Theorem 8 *If N is a normal subgroup of G with G/N a p-group and V is an indecomposable kN-module then V^G is also indecomposable.*

Proof Since G/N is a p-group, G has a sequence of subgroups

$$G = G_0 \supset G_1 \supset \cdots \supset G_r = N$$

each of index p in the preceding one. Suppose that we have already established the theorem in the case that the quotient group has order p. It then follows that $V^{G_{r-1}}$ is indecomposable and then that so is $(V^{G_{r-1}})^{G_{r-1}}$, and so on. But by the transitivity of induction, that is, part (4) of Lemma 5, the result of these successive inductions is isomorphic with V^G so V^G is indecomposable.

We now assume that $|G:N| = p$ and choose $g \in G$ such that gN is a generator for G/N. Hence, we have a vector space direct sum

$$V^G = 1 \otimes V \oplus g \otimes V \oplus \cdots \oplus g^{p-1} \otimes V.$$

Now $g^i \otimes V$ is a module for $g^i N g^{-i}$ isomorphic with the conjugate module $g^i(V)$ by transport. Hence, $g^i \otimes V$ is an indecomposable kN-module. Suppose that two of these modules are isomorphic: $g^i(V) \cong g^j(V)$, $0 \leqslant i < j \leqslant p-1$. Hence, $V \cong g^{-i}(g^i(V)) \cong g^{-i}(g^j(V)) = g^{j-i}(V)$, and similarly, since $g^{j-i}N$ generates G/N, we have that $V, g(V), \ldots, g^{p-1}(V)$ are all isomorphic. Thus, either the p summands of $(V^G)_N$ are all isomorphic or no two of them are isomorphic.

Suppose that $(V^G)_N$ has p summands pairwise non-isomorphic and that W is a non-zero direct summand of V^G. Thus, $W \cong g(W)$, which is trivial as $g \in G$, so $W_N \cong g(W_N)$ as kN-modules. By the Krull–Schmidt theorem, W_N has a direct summand isomorphic with $g^i(V)$ for some i, $0 \leqslant i < p$. Hence,

$g(W_N)$ has a summand isomorphic with $g(g^i(V)) \cong g^{i+1}(V)$, that is, $W_N \cong g(W_N)$ has such a summand. Hence, again by the Krull–Schmidt theorem, W_N is a decomposition into indecomposable modules, has summands isomorphic with each of $V, g(V), \ldots, g^{p-1}(V)$ so dim $W_N \geqslant p$ dim $V =$ dim V^G. Thus $W = V^G$ and V^G is indecomposable.

Finally, suppose that all the modules $g^i(V)$ are isomorphic. We shall prove that End(V^G) is local and so have the desired indecomposability. Let T be the linear transformation induced on V^G by g. Hence, End(V^G) consists of the linear transformations on V^G which commute with T and with the action of N, that is, End(V^G) consists of the elements of End$((V^G)_N)$ commuting with T. However, conjugation by T leaves End$((V^G)_N)$ invariant. Indeed, suppose S is a kN-endomorphism of V^G. If $u \in V^G$ and $y \in N$ then

$$
\begin{aligned}
TST^{-1}(yu) &= gSg^{-1}yu \\
&= gS(g^{-1}yg)g^{-1}u \\
&= gg^{-1}ygSg^{-1}u \\
&= y(TST^{-1})u.
\end{aligned}
$$

Hence, conjugation by T induces an automorphism of End$((V^G)_N)$ and the fixed points of this automorphism are the elements of End(V^G).

We have the direct decomposition

$$(V^G)_N = 1 \otimes V + \cdots + g^{p-1} \otimes V$$

of $(V^G)_N$ as a direct sum of kN-modules each isomorphic with V so End$((V^G)_N) \cong M_p(\text{End}(V))$. Let E_i, $1 \leqslant i \leqslant p$, be the endomorphism which is the identity on $g^{i-1} \otimes V$ and zero on all the other summands. Under the isomorphism just given, E_i is mapped to the p by p matrix I_i all of whose entries are zero except for the ith diagonal entry which is the identity endomorphism I of V. Since $T(g^i \otimes v) = g^{i+1} \otimes V$ it follows that $TE_iT^{-1} = E_{i+1}$ (where E_{p+1} denotes E_1) – remember these are all linear transformations of V^G. Hence, the automorphism α of $M_p(\text{End}(V))$, corresponding to the automorphism induced by T on End$((V^G)_N)$, permutes $I_1, I_2, \ldots, I_p$ cyclically. We have to see that the fixed-point of this automorphism in $M_p(\text{End}(V))$ is a local algebra.

Let $E = \text{End}(V)$ and $R = \text{rad } E$ so $M_p(R)$ is an ideal of $M_p(E)$ and it is also nilpotent: if $R^r = 0$ then any product of r elements from $M_p(R)$ is also zero. But $E/R \cong k$ so $M_p(E)/M_p(R) \cong M_p(k)$, a simple algebra, and so $M_p(R)$ is the radical of $M_p(E)$. Therefore, α leaves $M_p(R)$ invariant and so induces an automorphism β of $M_p(k)$. It is sufficient to prove that the algebra of fixed-points of β is local. Indeed, the algebra of fixed-points of α has nilpotent ideal (its intersection with $M_p(R)$) whose quotient is a subalgebra of the algebra of fixed-points of β and so this quotient is also local. But it is easy to see that if the quotient of an algebra by a nilpotent ideal is local then so is the algebra.

However, every automorphism of $M_p(k)$ is inner, as we saw in an early exercise. Hence, β is given by conjugation by a matrix M and this conjugation permuts cyclically the images of I_i in $M_p(k)$, that is, the p matrices all of whose entries are zero except for one diagonal entry which is 1. This forces, by a clear calculation,

$$
M = \begin{pmatrix}
0 & \lambda_1 & & & \\
 & 0 & \lambda_2 & & \\
 & & 0 & & \\
 & & & \ddots & \\
 & & & & \lambda_{p-1} \\
\lambda_p & & & & 0
\end{pmatrix}
$$

where $\lambda_1 \lambda_2 \cdots \lambda_p \neq 0$. Let $\mu \in k$ with $\mu^p = \lambda_1 \cdots \lambda_p$ so M is similar, by Jordan canonical form, to the matrix

$$
\begin{pmatrix}
\mu & 1 & & & \\
 & \mu & 1 & & \\
 & & \ddots & & \\
 & & & \mu & 1 \\
 & & & & \mu
\end{pmatrix}
$$

the p by p Jordan block for eigenvalue μ. However, the algebra of all matrices which commute with this matrix consists of all matrices of the form

$$
\begin{pmatrix}
\alpha_1 & \alpha_2 & & \alpha_p \\
 & \alpha_1 & \cdot & \\
 & & \cdot & \alpha_2 \\
 & & & \alpha_1
\end{pmatrix}
$$

which we have seen is local before. The proof is now complete.

Exercises

1 Verify the assertion on Jordan form made in the text.
2 (Mackey's Tensor Product Theorem.) Let H and L be subgroups of G and let U and V be a kH-module and a kL-module. Prove that

$$
U^G \otimes V^G \cong \bigoplus_{s \in H\backslash G/L} (s(U)_{L \cap sHs^{-1}} \otimes V_{L \cap sHs^{-1}})^G.
$$

3 With the notation of Lemma 6, let $\gamma \in \mathrm{Hom}_{kH}(U_H, V)$ and let $\gamma' \in \mathrm{Hom}_{kG}(U, V^G)$ be the corresponding map described in part (3) of Lemma 6. Show that $\pi\gamma' = \gamma$ where π is the homomorphism of V^G to V which maps $1 \otimes v$ to v and $s \otimes v$ to zero if $v \in V$, $s \in G$, $s \notin H$.

9 Vertices and sources

We have just generalized the notion of free modules and we shall now do the same for projective modules. This will allow us to establish the first connection between kG-modules and modules for p-subgroups.

Proposition 1 *If U is a kG-module and H is a subgroup of G then the following are equivalent:*

(1) *U is a direct summand of a relatively H-free module;*

(2) *If φ is a homomorphism of the kG-module V onto U and φ is split as a kH-homomorphism then φ is split;*

(3) *If φ is a homomorphism of the kG-module V onto the kG-module W and ψ is a homomorphism of U to W then there is a kG-homomorphism ρ of U to V with $\varphi\rho = \psi$ provided there is a kH-homomorphism with this property;*

(4) *U is a direct summand of $(U_H)^G$.*

The first three statements are easily seen to be equivalent; the proof of Theorem 5.2 carries over immediately. One need only use relatively free modules in place of free modules and use the existence of kH-homomorphisms where one had used linear transformations. As for the fourth statement, it clearly implies the first, being a refined version of it. On the other hand, there is a kG-homomorphism of $(U_H)^G$ onto U which is split as a kH-homomorphism: the identity map of U_H to U extends (after the usual identification $U = 1 \otimes U$) to a kG-homomorphism of $(U_H)^G$ to U and this splits over H as the kH-homomorphism of U to $(U_H)^G$, sending $u \in U$ to $1 \otimes u$, makes clear. Hence, if (2) holds then so does (4).

The modules satisfying the properties of the proposition are called *relatively H-projective*. We now investigate the question of for which subgroups H is a kG-module relatively H-projective; the rest of the section revolves around this point.

Theorem 2 *If H is a subgroup of G containing a Sylow p-subgroup of G then every kG-module is relatively H-projective.*

We already have proved the case that $H = 1$ and this is another generalıza-
tion which replaces the identity subgroup with an arbitrary one. Indeed, if
the hypothesis applies with $H = 1$ this means that G has order not divisible
by p so every kG-module is semisimple and so every kG-module is
projective.

The proof is also an easy extension of the argur.ent establishing Theorem
3.1. Indeed, suppose that φ is a homomorphism of the kG-module V onto
the kG-module U and φ splits as a kH-homomorphism: we must show that
φ splits as a kG-homomorphism so we can invoke Proposition 2. We have
that the kernel W is a summand of V when considered as a kH-submodule
and we must prove that it is a summand as a kG-submodule. Let π be a kH-
homomorphism of V onto W which is the identity on W and zero on a kH-
submodule which together with W gives a direct sum decomposition of V.
Set, as in Theorem 3.1,

$$\pi' = \frac{1}{|G:H|} \sum_{s \in G/H} s\pi s^{-1}$$

where the element s also denotes the corresponding linear transformation
of V. Note that we can divide by $|G:H|$ as this index is not divisible by p by
hypothesis. It is now easy to check, just as in Theorem 3.1, that π' is a kG-
homomorphism, $\pi'(V) = W$ and π' is the identity on W so V is the direct sum
of W and the kG-module which is the kernel of π' and the theorem is proved.

Corollary 3 *If H is a subgroup of G containing a Sylow p-subgroup and U is
a kG-module with U_H projective then U is projective.*

Proof We have that U is relatively H-projective so U is a direct summand
of $(U_H)^G$. However, U_H is projective so $(U_H)^G$ is also, by the Omnibus
Lemma of the preceding section.

Since we are going to be using direct decomposition extensively, we
introduce the standard notation: if U and V are kG-modules and U is
(isomorphic with) a direct summand of V then we write $U \mid V$.

Theorem 4 *Let U be an indecomposable kG-module.*

(1) *There is a p-subgroup Q of G, unique up to conjugacy in G, such that U
is relatively H-projective, for a subgroup H of G, if, and only if, H contains a
conjugate of Q.*

(2) *There is an indecomposable kQ-module S, unique up to conjugacy in
$N_G(Q)$, such that $U \mid S^G$.*

The subgroup Q is a *vertex* of U and the module S is a *source* of U. The idea is that the closer Q is to the identity, the nearer U is to being projective. Indeed, U is projective if, and only if, $Q = 1$. For if U is projective then U is a summand of a free module, that is, a module induced from the identity subgroup, so $Q = 1$ by (1), while if $Q = 1$ then (2) shows that U is projective, as does (1).

Proof Since U is relatively projective for any Sylow p-subgroup we can choose a p-subgroup Q of smallest possible order such that U is relatively Q-projective. In particular, $U \mid (U_Q)^G$ so there is an indecomposable summand S of U_Q with $U \mid S^G$. If H is a subgroup of G containing Q then $U \mid (S^H)^G$, as $(S^H)^G \cong S^G$, so U is also relatively H-projective. This also implies that U is relatively gHg^{-1}-projective for any $g \in G$. Indeed, $U \cong g(U)$ and $g(U)$ is relatively gHg^{-1}-projective, by transport of structure.

On the other hand, suppose that H is a subgroup of G with U relatively H-projective and that V is an indecomposable kH-module with $U \mid V^G$. We shall prove that H contains a conjugate of Q and that if $H = Q$ then V is a conjugate of S by an element of $N_G(Q)$. Since $U \mid S^G$ we trivially have $U \mid g(S)^G$, for any $g \in G$, and if $g \in N_G(Q)$ then $g(S)$ is a kQ-module; this will complete the proof of the theorem.

But $S \mid U_Q$ and $U \mid V^G$ so $S \mid (V^G)_Q$ which, by Mackey's theorem, can be expressed,

$$(V^G)_Q \cong \bigoplus_{s \in Q \backslash G / H} (s(V)_{Q \cap sHs^{-1}})^Q$$

Hence, by the Krull–Schmidt theorem, $S \mid (s(V)_{Q \cap sHs^{-1}})^Q$, for some $s \in G$. But this implies that $U \mid (s(V)_{Q \cap sHs^{-1}})^G$, as $U \mid S^G$, so our choice of Q forces $Q = Q \cap sHs^{-1}$, that is, $Q \subseteq sHs^{-1}$ or $s^{-1}Qs \subseteq H$ as claimed. Moreover, if $Q = H$ then $Q \cap sQs^{-1} = Q$ means $s \in N(Q)$ and so $S \cong s(V)$ and $V \cong s^{-1}(S)$ as asserted.

Let us see what this means for a non-projective module. Let P be a Sylow p-subgroup of G and choose a vertex Q of the trivial kG-module k. Since k_Q is a trivial kQ-module it is indecomposable so k_Q is a source of k and $k \mid (k_Q)^G$ Hence,

$$k_P \mid ((k_Q)^G)_P \cong \bigoplus_{s \in P \backslash G / Q} (s(k_Q)_{sQs^{-1} \cap P})^P$$

so $k_P \mid (k_{sQs^{-1} \cap P})^P$ for some $s \in G$. Set $R = sQs^{-1} \cap P$: we claim that $(k_R)^P$ is indecomposable. This will imply that $R = P$, that is, $P = sQs^{-1}$ so Q is a

Sylow p-subgroup of G. Note that it also shows that P is conjugate with Q, that is, re-proves the conjugacy of the Sylow p-subgroups of G!

Now we shall prove that the socle of $(k_R)^P$ is one-dimensional, so $(k_R)^P$ is certainly indecomposable. But the only simple kP-module is k so the socle has dimension equal to the dimension of $\mathrm{Hom}_{kP}(k, (k_R)^P)$. However,

$$\mathrm{Hom}_{kP}(k, (k_R)^P) \cong \mathrm{Hom}_{kR}(k_R, k_R)$$

is one-dimensional and all our assertions are established.

The rest of the section is devoted to a number of useful results about vertices. This will give the reader a good opportunity to absorb these ideas through using them.

Lemma 5 *If U is an indecomposable kG-module with vertex Q and H is a subgroup containing Q then there is an indecomposable kH-module V satisfying any two of the following statements:*

(1) $V \mid U_H$;

(2) $U \mid V^G$;

(3) V has vertex Q.

Hence, we are producing three modules and so are really proving three lemmas. The reason that we have put these results together is that one can find a module which simultaneously satisfies all three assertions but the proof of this depends on the Green correspondence which in turn relies on this lemma.

Proof Since U is relatively H-projective, we have $U \mid (U_H)^G$ and therefore there is an indecomposable summand V of U_H such that $U \mid V^G$. Hence, V satisfies (1) and (2).

Next, let S be a kQ-module which is a source for U so $U \mid S^G$. But $S^G \mid (S^H)^G$ so there is an indecomposable summand V of S^H with $U \mid V^G$. We claim that V has vertex Q. Once we establish this assertion we will have that V satisfies (2) and (3). Since $V \mid S^H$ we have that V is relatively Q-projective so there is a vertex R of V contained in Q. Let W be a kR-module such that $V \mid W^H$. Hence, $V^G \mid (W^H)^G$, that is, $V^G \mid W^G$, so $U \mid W^G$ and U is also relatively R-projective. Thus, R contains a conjugate of Q. But $R \subseteq Q$ so $R = Q$ just as claimed.

As in the proof of Theorem 4, let S be an indecomposable kQ-module with $S \mid U_Q$ and $U \mid S^G$. Hence, there is an indecomposable kH-module V with $V \mid U_H$ and $S \mid V_Q$. We shall prove that V has vertex Q and so fulfills conditions (1) and (3).

But $V \mid U_H$ so $V \mid (S^G)_H$ and by Mackey's theorem there is $s \in G$ with

$$V \mid (s(S)_{H \cap sQs^{-1}})^H.$$

Hence, V has a vertex R with $R \subseteq H \cap sQs^{-1}$. It suffices to prove that R is conjugate to Q in H. However, V is a summand of a module induced from R to H and $S \mid V_Q$ so Mackey's theorem implies that S is relatively $Q \cap hRh^{-1}$ projective for some h in H. But S has vertex Q (or else, as $U \mid S^G$, U would be relatively projective for a proper subgroup of Q) so $Q \cap hRh^{-1}$ cannot be a proper subgroup of Q, that is, $Q \subseteq hRh^{-1}$. But $R \subseteq sQs^{-1}$ so $|R| \leqslant |Q|$ and so we must have $Q = hRh^{-1}$.

Lemma 6 *If V is a relatively Q-projective kH-module, where Q is a subgroup of the subgroup H of G, then*

$$(V^G)_H \cong V \oplus W$$

where every indecomposable summand of W is relatively projective for a subgroup of the form $sQs^{-1} \cap H$, $s \in G$, $s \notin H$.

Proof Since V is relatively Q-projective, there is a kQ-module U with $V \mid U^H$. Thus $U^H \cong V \oplus T$ for some kH-module T, so $U^G \cong V^G \oplus T^G$ and

$$(U^G)_H \cong V \oplus W \oplus T \oplus X$$

where $(V^G)_H \cong V \oplus W$ and $(T^G)_H \cong T \oplus X$ for suitable kH-modules W and X. However, by Mackey's theorem,

$$(U^G)_H \cong \bigoplus_{s \in H \backslash G / Q} (s(U)_{H \cap sQs^{-1}})^H \cong U^H \oplus Y$$

where the summand $s \in H$ gives U^H, Y is the direct sum of all terms for $s \notin H$ and so each indecomposable summand of Y is relatively projective for a subgroup of the form $sQs^{-1} \cap H$, $s \notin H$. The Krull–Schmidt theorem implies that $W \oplus X \cong Y$ so W is as claimed.

Lemma 7 *If U is an indecomposable kG-module with vertex Q and trivial source and H is any subgroup of G then U_H has an indecomposable summand with a vertex containing $Q \cap H$.*

Proof Since U has vertex Q and source k_Q we have, in particular, that $k_Q \mid U_Q$ so that $k_{Q \cap H} \mid U_{Q \cap H}$. Hence, there is an indecomposable summand V of U_H with $k_{Q \cap H} \mid V_{Q \cap H}$. Let R be a vertex of V; we shall prove that a conjugate of R in H contains $Q \cap H$ and so establish the lemma.

Since V is relatively R-projective, Mackey's theorem immediately implies that every indecomposable summand of $V_{Q \cap H}$ is relatively projective for a subgroup of the form $(Q \cap H) \cap hRh^{-1}$ for some h in H. But $k_{Q \cap H}$ is a summand of $V_{Q \cap H}$ and $k_{Q \cap H}$ has vertex $Q \cap H$ (as $Q \cap H$ is its own Sylow p-subgroup) so a conjugate of $Q \cap H$ in H is contained in hRh^{-1} for some h in H, as claimed.

Lemma 8 *Let U be a kG-module such that U_N is indecomposable, where N is a normal subgroup of G. If Q is a vertex of U then QN/N is a Sylow p-subgroup of G/N.*

Proof Let S be a Sylow p-subgroup of G which contains Q so SN/N is a Sylow p-subgroup of G/N containing QN/N. By Lemma 5, there is an indecomposable summand of U_{SN} which has vertex Q. However, U_N is indecomposable so certainly U_{SN} is also and we have, in particular, that U_{SN} is relatively QN-projective and so $U_{SN} \mid (U_{QN})^{SN}$. Since QN/N is a subgroup of the p-group SN/N there is a series of subgroups connecting them such that each term of the series is normal in the preceding one. Hence, by iterating use of Green's Indecomposability Criterion, we have that $(U_{QN})^{SN}$ is indecomposable so $U_{SN} \cong (U_{QN})^{SN}$. Hence, as $\dim((U_{QN})^{SN}) = |SN:QN| \dim U$, we have $SN = QN$ as claimed.

Exercises

1   Generalize the notion of the injective modules to relatively injective modules and state and prove a result for such modules analogous to Proposition 1.

2   If U and V are kG-modules and U is relatively H-projective for a subgroup H of G then so is $U \otimes V$. (Hint: Do the same for relatively free modules.)

3   If V is an indecomposable kQ-module, where Q is a p-subgroup of G, and $\dim V$ is not divisible by p then V has vertex Q.

4   (cont.) There is an indecomposable summand of V^G with vertex Q.

5   (cont.) Deduce that Q is contained in a Sylow p-subgroup of G.

6   If H is a subgroup of G and U is an indecomposable kG-module with $U \mid (k_H)^G$ then U has a trivial source. (Hint: Let Q be a vertex of U and apply Mackey's theorem to $((k_H)^G)_Q$.)

10 Trivial intersections

In this section we shall look at a special case of the Green correspondence, which is the definitive result connecting indecomposable kG-modules with modules for local subgroups. The main reason for doing this is that the key ideas are much more accessible in this case. Moreover, we shall be able to redo the results on projective modules for SL(2, p) in a way that is illustrative of things to come.

We fix now a Sylow p-subgroup P of G and *assume* for the remainder of the section that P is a trivial intersection subgroup, that is, whenever $g \in G$ then $P \cap gPg^{-1}$ is either P or 1. Let $L = N(P)$ so if $g \in G$, $g \notin L$ then $P \cap gPg^{-1} = 1$. The main result of this section can now be stated.

Theorem 1 *There is a one-to-one correspondence between isomorphism classes of non-projective indecomposable kG and kL-modules, such that if U and V are such modules, for G and H respectively, then*

$$U_L \cong V \oplus Q$$
$$V^G \cong U \oplus R$$

where Q and R are projective kG and kL-modules, respectively.

Thus, the correspondence is determined by either restriction or induction. Indeed, U_H has exactly one non-projective indecomposable summand, up to isomorphism, and a similar thing can be said about V^G.

Proof By Mackey's theorem, we have

$$(V^G)_L \cong \bigoplus_{s \in L \backslash G / L} (s(V)_{L \cap sLs^{-1}})^L$$

so $(V^G)_L$ is isomorphic with the direct sum of V (the summand corresponding to $s \in L$) and modules induced from subgroups $L \cap sLs^{-1}$, where $s \notin L$. However, P and sPs^{-1} are the unique Sylow p-subgroups of L and sLs^{-1} so that $P \cap sPs^{-1}$ is the Sylow p-subgroup of $L \cap sLs^{-1}$. Hence, if $s \notin L$ then our assumption yields that $L \cap sLs^{-1}$ has order not divisible by p and so every module for $L \cap sLs^{-1}$ is semisimple so every module is projective (as all short exact sequences must split). Thus, we have $(V^G)_L \cong V \oplus Y$, where Y is a projective kL-module, since projectives induce to projectives.

Express $V^G \cong U_1 \oplus \cdots \oplus U_n$, where each U_i is indecomposable. Since L contains a Sylow p-subgroup of G, we have that U_i is projective if, and only if, $(U_i)_L$ is projective. However, $(V^G)_L$ has a unique non-projective indecomposable summand in any decomposition into the direct sum of indecomposable modules so it must be that all the U_i, with exactly one exception, are projective. Choose notation so that $U = U_1$ is not projective so we now have $V^G \cong U \oplus Q$, where Q is projective and also that $U_L \cong V \oplus R$, where R is projective, since $V \oplus Y \cong (V^G)_L \cong U_L \oplus Q_L$.

Moreover, suppose that U is any non-projective indecomposable kG-

module. It follows that U arises as in the above argument. Indeed, L contains a Sylow p-subgroup of G so every kG-module is relatively L-projective. In particular, there is an indecomposable kL-module V such that $U \mid V^G$, and V must be non-projective as V^G is non-projective. Thus, by the above arguments we have $U_L \cong V \oplus R$, for some projective kL-module R. We therefore have a map from isomorphism classes of non-projective indecomposable kG-modules to such classes of kL-modules as well as a map in the other direction constructed above. Since we have already shown that the compositions of these two maps, in any order, are identities the theorem is fully proved.

This correspondence has a number of other important properties, one of which is the following result.

Corollary 2 *If U_1 and U_2 are non-projective indecomposable kG-modules while V_1 and V_2 are the corresponding kL-modules then there is a non-split exact sequence*

$$0 \to U_1 \to U \to U_2 \to 0$$

if, and only if, there is a non-split exact sequence

$$0 \to V_1 \to V \to V_2 \to 0.$$

Proof First, suppose that

$$0 \to V_1 \to V \to V_2 \to 0$$

is non-split so that so is

$$0 \to V_1^G \to V^G \to V_2^G \to 0$$

by the Omnibus Lemma of section 8. Express $V_i^G \cong U_i \oplus R_i$, $i = 1, 2$, where R_i is a projective kG-module.

For the sake of simplicity, and since there is no loss of generality, we make the following identifications: $V_1^G = U_1 + R_1$, $V_1 \subseteq V$ (so $V_1^G \subseteq V^G$). Hence, there is a submodule W of V^G containing V_1^G such that $V^G/W \cong R_2$, $W/V_1^G \cong U_2$. This gives the following picture:

Hence, we have a short exact sequence

$$0 \to U_1 \to W/R_1 \to U_2 \to 0$$

and we claim this does not split. Indeed, assume the contrary; we shall show that V_1^G is a direct summand of V^G. Our assumption means there is a submodule X such that $W = X + V_1^G$ while $X \cap V_1^G = R_1$. But R_1 is also injective so X is a direct sum: $X = Y + R_1$. Hence, $W = Y + V_1^G$ is a direct sum. (We have $Y + V_1^G = Y + R_1 + V_1^G = X + V_1^G = W$ and $Y \cap V_1^G = Y \cap X \cap V_1^G = Y \cap R_1 = 0$.) Thus, we have the picture

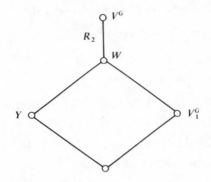

But R_2 is projective so V^G/Y has W/Y as a direct summand, that is, there is a submodule Z with $V^G = Z + W$ and $Z \cap W = Y$. Thus, $V^G = Z + V_1^G$ is a direct sum (as $Z + V_1^G = Z + Y = V_1^G = Z + W = V_1^G$ and $Z \cap V_1^G = Z \cap W \cap V_1^G = Y \cap V_1^G = 0$).

Conversely, suppose that

$$0 \to U_1 \to U \to U_2 \to 0$$

is not split so

$$0 \to (U_1)_L \to U_L \to (U_2)_L \to 0$$

is not either: the module U_2 is relatively L-projective as L contains a Sylow p-subgroup so splitting for kL-modules implies splitting. Since $U_i \cong V_i \oplus Q_i$, where Q_i is projective, we can proceed just as above, to produce a short exact sequence of the desired sort which does not split because if it did so would the sequence just given involving U_L. This proves the result.

We shall now give one more property of our correspondence, before specializing to SL(2, p). To state this result, we need one definition (which makes sense for arbitrary G). If U_1 and U_2 are kG-modules then

$$\overline{\mathrm{Hom}}_{kG}(U_1, U_2)$$

is the quotient of the vector space $\text{Hom}_{kG}(U_1, U_2)$ by the subspace of all homomorphisms of U_1 to U_2 which factor through a projective kG-module (that is, are the composition of a homomorphism of U_1 to a projective with a homomorphism to U_2). This is indeed a subspace of $\text{Hom}_{kG}(U_2, U_2)$. (If $f_1, f_2 \in \text{Hom}_{kG}(U_1, U_2)$ factor through the projective kG-modules P_1 and P_2, respectively, then it is easy to see that $f_1 + f_2$ factors through $P_1 \oplus P_2$, while λf_1, for any scalar λ, also factors through P_1.)

Theorem 3 *If U_1 and U_2 are non-projective indecomposable kG-modules and V_1 and V_2 are the corresponding kL-modules then*

$$\overline{\text{Hom}}_{kG}(U_1, U_2) \cong \overline{\text{Hom}}_{kL}(V_1, V_2).$$

Proof We shall show that

$$\overline{\text{Hom}}_{kG}(V_1^G, U_2) \cong \overline{\text{Hom}}_{kL}(V_1, (U_2)_L)$$

and this is sufficient. Indeed, there are suitable projective modules R_1 and Q_2 with $V_1^G \cong U_1 \oplus R_1$, $(U_2)_L \cong V_2 \oplus Q_2$ so then

$$\overline{\text{Hom}}_{kG}(V_1^G, U_2) \cong \overline{\text{Hom}}_{kG}(U_1 \oplus R_1, U_2) \cong \overline{\text{Hom}}_{kG}(U_1, U_2)$$

(the last step is easy to check) and similarly

$$\overline{\text{Hom}}_{kL}(V_1, (U_2)_L) \cong \overline{\text{Hom}}_{kL}(V_1, V_2 \oplus Q_2) \cong \overline{\text{Hom}}_{kL}(V_1, V_2).$$

But if U is a kG-module and V is a kL-module, we already know that

$$\text{Hom}_{kG}(V^G, U) \cong \text{Hom}_{kL}(V, U_L)$$

in an explicit way so we need only show here that for two maps which correspond under this isomorphism, one factors through a projective if, and only if, the other one does.

Suppose that $\gamma \in \text{Hom}_{kL}(V, U_L)$ and $\hat{\gamma}$ is the corresponding map in $\text{Hom}_{kG}(V^G, U)$ so if we regard V as contained in V^G (by identification with $1 \otimes V$) then $\hat{\gamma}$ is an extension of γ. In particular, if $\hat{\gamma}$ factors through the projective kG-module R then certainly γ factors through R_L which is also projective. On the other hand, suppose $\gamma = \beta\alpha$, where $\alpha \in \text{Hom}_{kL}(V, Q)$, $\beta \in \text{Hom}_{kL}(Q, U_L)$ and Q is a projective kL-module, which is pictured as follows.

$$
\begin{array}{ccc}
 & Q & \\
{\scriptstyle\alpha}\nearrow & & \searrow{\scriptstyle\beta} \\
V & \xrightarrow{\hphantom{xx}\gamma\hphantom{xx}} & U_L
\end{array}
$$

Let α^G be the homomorphism of V^G to Q^G which is the extension of the homomorphism α of V to Q and let $\hat{\beta}$ be the homomorphism of Q^G to U which corresponds to β so it is also an extension of β. Thus, $\hat{\beta}\sigma^G$ is a kG-

homomorphism of V^G to U and on V it coincides with $\beta\alpha = \gamma$. Hence, by the uniqueness property of induced modules, $\hat{\beta}\alpha^G = \hat{\gamma}$ and since Q^G is projective we also have that $\hat{\gamma}$ factors through a projective module.

For the remainder of the section we set $G = \mathrm{SL}(2, p)$ and let P be the set of matrices of the form

$$\begin{pmatrix} 1 & 0 \\ c & 1 \end{pmatrix}$$

so that P is a Sylow p-subgroup of G (and is certainly a trivial intersection subgroup as it has order p). If $L = N(P)$ then L consists of all matrices of the form

$$\begin{pmatrix} a & 0 \\ c & d \end{pmatrix}$$

with $ad = 1$. The map which attaches to such an element of L an element a of the prime field is a homomorphism and it has kernel P. Therefore, L/P is cyclic of order $p-1$ and so kL has $p-1$ simple modules each being of dimension one. Since P has order p and is normal in L we also know that all the indecomposable kL-modules are uniserial.

If j is any integer, let S_j be the one-dimensional kL-module in which the above element of L acts by multiplication by a^j. There is certainly such a module since the map which attaches a^j to the given element is a homomorphism. Since L/P is cyclic of order $p-1$ we have that S_{j_1} and S_{j_2} are isomorphic if, and only if, $j_1 \equiv j_2$ (modulo $p-1$). Moreover, it is also clear that $S_{j_1} \otimes S_{j_2} \cong S_{j_1+j_2}$.

Let us now consider V_2 as a kL-module. Recall that V_2 has a basis consisting of X and Y and that the above element of L maps X to $aX + cY$ and Y to $a^{-1}Y$ (as $a^{-1} = d$). Thus kY is a submodule of $(V_2)_L$ isomorphic with S_{-1} and its quotient is isomorphic with S_1, as the image of X is a basis. Moreover, $(V_2)_L$ is uniserial. Indeed, we would otherwise have that $(V_2)_L$ has another maximal submodule other than kY and this would have to be isomorphic with S_1. However, it is easy to check that there is no such submodule. Thus, we may picture the full lattice of submodules of $(V_2)_L$ as follows

$$
\begin{array}{c}
\circ \\
| \quad S_1 \\
\circ \\
| \quad S_{-1} \\
\circ
\end{array}
$$

This reveals the detailed structure of all the indecomposable kL-modules by telling us which composition factors appear and in which order. Indeed, we know that there is a simple kL-module W such that whenever S_j is a composition factor of an indecomposable module then the 'next' composition factor is $S_j \otimes W$. However, the simple kL-modules are $S_{-1}, S_{-2}, \ldots, S_{-(p-1)}$ and their tensor products with S_1 are $S_0, S_{1-1}, \ldots, S_{-(p-2)}$, which again are the distinct simple kL-modules, so we must have $W \cong S_{-2}$. Thus, if U is an indecomposable kL-module and $U/\text{rad}(U) \cong S_j$ then the composition factors in the composition series of U are $S_j, S_{j-2}, S_{j-4}, \ldots$ and so on, depending only on dim U, which is at most p.

We can now determine easily the structure of $(V_i)_L$, $1 \leqslant i < p$. In our proof of the simplicity of these modules we showed that the socle of $(V_i)^P$, that is, the fixed-points, was one-dimensional and had Y^{i-1} as a basis. Since P acts trivially on each simple kL-module we have that kY^{i-1} is the socle of $(V_i)_L$ and, by the formula for the action in terms of X and Y, we have this socle is isomorphic with $S_{-(i-1)}$. Since V_i is of dimension i, we can picture it as follows.

It is the i-dimensional indecomposable with radical quotient S_{i-1} as $(i-1) - 2(i-1) = -i+1$.

We can now start to describe the structure of kG-modules.

Lemma 4 *There is a non-split short exact sequence of kG-modules, $1 \leqslant i < p-1$,*

$$0 \to V_{p-i-1} \to V \to V_i \to 0.$$

Proof Let U be the indecomposable kL-module of dimension $p-1$ with $U/\text{rad}(U) \cong S_{i-1}$. Thus $U/\text{rad}^i(U)$ is i-dimensional and also has radical quotient isomorphic with S_{i-1}. Therefore $U/\text{rad}^i(U) \cong (V_i)_L$. The radical quotient of $\text{rad}^i(U)$ is the simple module 'next' after the socle of $U/\text{rad}^i(U)$, that is, S_{-i+1}, so it is $S_{-i-1} \cong S_{(p-1-i)-1}$. But $\mathrm{rad}^i(U)$ is of dimension $(p-1)-i$ so we must also have $\mathrm{rad}^i(U) \cong (V_{p-1-i})_L$. Hence, we have a short exact sequence

$$0 \to (V_{p-1-i})_L \to U \to (V_i)_L \to 0$$

and this cannot split because U is indecomposable. Therefore, the lemma is a consequence of Corollary 2.

Lemma 5 *There is a non-split short exact sequence of kG-modules*, $1 \leqslant i \leqslant p-1$,

$$0 \to V_{p+1-i} \to V \to V_i \to 0.$$

Proof Set $U = Q_{i-1} \oplus S_{-i+1}$ where Q_{i-1} is the indecomposable projective kL-module corresponding with S_{i-1} so Q_{i-1} has dimension p, is uniserial and has radical quotient isomorphic with S_{i-1}. Set $W = \mathrm{rad}^i(Q_{i-1}) \subseteq U$ so W is $(p-i)$-dimensional and $W/\mathrm{rad}(W) \cong S_{-i+1}$ (as $i-1-2(i-1) = -i+1$). Thus, $W \cong (V_{p+1-i})_L$ as the latter module has radical quotient S_{p-i} and $p-i \equiv -i+1$ (modulo $p-1$). Let φ be a homomorphism of W onto S_{-i+1} and let Z be the submodule of $W \oplus S_{-i+1}$ consisting of all $(w, \varphi(w))$ as w runs over W. Projection on the first component shows that $W \cong Z$. We claim that $U/Z \cong (V_i)_L$.

Certainly, U/Z is i-dimensional as W and Z are of dimension $p+1-i$. Hence, it is enough to show that the radical quotient of U/Z is isomorphic with S_{i-1}. However, $\mathrm{rad}(U/Z) = (\mathrm{rad}(U) + Z)/Z$; hence we need only prove that $\mathrm{rad}(U) + Z \supseteq \mathrm{rad}(Q_{i-1}) \oplus S_{-i+1}$ as $\mathrm{rad}(U) + Z \subset U$ and the quotient of $Q_{i-1} \oplus S_{-i+1}$ by $\mathrm{rad}(Q_{i-1}) \oplus S_{-i+1}$ is isomorphic with S_{i-1}. But certainly $\mathrm{rad}(Q_{i-1}) \subseteq \mathrm{rad}\, U$ so it suffices to show that $S_{-i+1} \subseteq \mathrm{rad}(U) + Z$. If $s \in S_{-i+1}$ choose $w \in W$ with $\varphi(w) = s$. Thus, $(0, s) = (w, s) - (w, 0) \in Z + W$. But $W = \mathrm{rad}^i(Q_{i-1}) \subseteq \mathrm{rad}(U)$ so U/Z is as claimed.

Hence, we have a short exact sequence

$$0 \to (V_{p+1-i})_L \to U \to (V_i)_L \to 0$$

and this cannot split: for $U = Q_{i-1} \oplus S_{-i+1}$ so the Krull–Schmidt theorem guarantees that U is not isomorphic with $(V_{p+1-i})_L \oplus (V_i)_L$. The lemma now follows immediately from Corollary 2.

We are now prepared to analyze the structure of the indecomposable projective kG-modules $P_1, \ldots, P_p$, $p > 2$. First, Lemma 4 guarantees the existence of a uniserial module with composition factors V_1, V_{p-2} so this must be a homomorphic image of P_1 and so we have a picture of P_1 as

follows

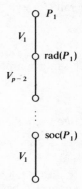

Thus, dim $P_1 \geqslant 1 + (p-2) + 1 = p$ and we have equality if, and only if, $\text{rad}(P_1)/\text{soc}(P_1) \cong V_{p-2}$. Similarly, using Lemma 5, we have that dim $P_{p-1} \geqslant 2p$ and that equality holds if, and only if, $\text{rad}(P_{p-1})/\text{soc}(P_{p-1}) \cong V_2$. Next, suppose that $1 < i < p-1$. We can apply both Lemma 4 and Lemma 5 and we deduce that P_i has submodules U_+ and U_- of $\text{rad}(P_i)$ such that $\text{rad}(P_i)/U_+ \cong V_{p+1-i}$ and $\text{rad}(P_i)/U_- \cong V_{p-1-i}$. Therefore, if $U = U_+ \cap U_-$ then $\text{rad}(P_i)/U \cong V_{p+1-i} \oplus V_{p-1-i}$ and we have also a picture

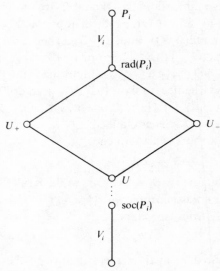

Hence, dim $P_i \geqslant i + (p+1-i) + (p-1-i) + i = 2p$ and equality holds if, and only if, $\text{rad}(P_i)/\text{soc}(P_i) \cong V_{p+1-i} \oplus V_{p-1-i}$.

Moreover, we also have

$$p(p^2 - 1) = \dim kG$$

$$= \sum_{i=1}^{p} \dim V_i \dim P_i$$

$$\geqslant p + (2 + \cdots + (p-2)2p + (p-1)2p + p \cdot p$$

$$= p^3 - p.$$

Thus, we must have equalities throughout and so all the P_i are as proved before.

We want to point out that this proof gives even more. We did not have to assume that $V_1, \ldots, V_p$ were the only simple modules or that V_p was projective! Indeed, in the last calculation we would instead have

$$p(p^2 - 1) = \dim G \geqslant \sum_{i=1}^{p} (\dim V_i)(\dim P_i)$$

as there might be terms corresponding to other projective modules. Hence, we deduce there are no others, so $V_1, \ldots, V_p$ are the only simple modules, and that $\dim P_p = p = \dim V_p$ so $V_p \cong P_p$ is projective.

Exercises

1 Let φ be a homomorphism of the projective kG-module Q onto the kG-module V. If ψ is a homomorphism of the kG-module U to V show that ψ factors through a projective if, and only if, there is a homomorphism α of U to Q such that $\varphi\alpha = \psi$.

2 With the notation of Theorem 3, let $\varphi \in \mathrm{Hom}_{kG}(U_1, U_2)$ and show that φ factors through a projective if φ does when considered as an element of $\mathrm{Hom}_{kL}((U_1)_L, (U_2)_L)$.

3 (cont.) Let i be a homomorphism of V_1 into a direct summand of U_1 and let π be a homomorphism of U_2 onto V_2 with kernel a direct summand. If $\varphi \in \mathrm{Hom}_{kG}(U_1, U_2)$ then $\pi\varphi i \in \mathrm{Hom}_{kL}(V_1, V_2)$. Prove that the map sending φ to $\pi\varphi i$ induces an isomorphism of $\overline{\mathrm{Hom}_{kG}(U_1, U_2)}$ onto $\overline{\mathrm{Hom}_{kL}(V_1, V_2)}$. (Hint: use Theorem 3).

11 Green correspondence

The correspondence established in the last section is a special case of the Green correspondence which applies to all groups without any reference to the embedding of the Sylow p-subgroup. It is the fundamental theorem of this approach to representation theory. The general result is encumbered with a number of important technicalities so we shall begin by describing another special case.

Let Q be a p-subgroup of G and set $L = N_G(Q)$. The Green correspondence implies that there is a one-to-one correspondence between the isomorphism classes of indecomposable kG-modules with vertex Q and the isomorphism classes of such kL-modules. Moreover, this happens in a very nice manner. Indeed, let U be an indecomposable kG-module with vertex Q. We can now express

$$U_L \cong V \oplus Y$$

where V is indecomposable and has vertex Q and where no indecomposable summand of Y has vertex Q. Hence, by the Krull–Schmidt theorem, the isomorphism class of V is well determined. Similarly, if V is an indecomposable kL-module with vertex Q then

$$V^G \cong U \oplus X$$

where U is an indecomposable kG-module with vertex Q and no indecomposable summand of X has vertex Q. Finally, these two correspondences are inverse to each other.

In the last chapter, we dealt with a Sylow p-subgroup and all indecomposable non-projective modules for G and the normalizer of a Sylow p-subgroup, not just those with vertex the Sylow subgroup. The general result will include this feature: we will be able to deal with other kG-modules and other kL-modules, not just the ones with vertex Q.

Let us fix some notation. We let Q be a p-subgroup of G and L a subgroup containing $N_G(Q)$. If P and R are subgroups of G we write $P \subseteq_G R$ to mean that a conjugate of P in G is contained in R. If $\mathscr{H}$ is a collection of subgroups of G then $P \subseteq_G \mathscr{H}$ means $P \subseteq_G H$ for some $H \in \mathscr{H}$. We now fix some collections of p-subgroup of G. Let

$$\mathfrak{x} = \{sQs^{-1} \cap Q \mid s \in G, s \notin L\},$$
$$\mathfrak{y} = \{sQs^{-1} \cap L \mid s \in G, s \notin L\},$$
$$\mathfrak{z} = \{R \mid R \subseteq Q, R \not\subseteq_G \mathfrak{x}\}.$$

One should think of $\mathfrak{x}$ and $\mathfrak{y}$ as consisting of 'small' subgroups, with respect to Q (even though $\mathfrak{y}$ may well contain a conjugate of Q) and that $\mathfrak{z}$ consists of Q and its 'large' subgroups, the typical ones. Note that each subgroup of $\mathfrak{x}$ is a proper subgroup of Q, since $L \supseteq N_G(Q)$, so also $Q \in \mathfrak{z}$. Finally, if $\mathscr{H}$ is any collection of subgroups of G we shall say that the kG-module U is relatively

$\mathcal{H}$-projective if U is the direct sum of modules each of which is relatively projective for a subgroup of $\mathcal{H}$. We can now state the fundamental result.

Theorem 1 *There is a one-to-one correspondence between isomorphism classes of indecomposable kG-modules with vertex in $\mathfrak{z}$ and isomorphism classes of indecomposable kL-modules with vertex in $\mathfrak{z}$. If U and V are such modules for G and L, respectively, which correspond then U and V have the same vertex and*

$$U_L \cong V \oplus Y,$$
$$V^G \cong U \oplus X$$

where Y is a relatively $\mathfrak{y}$-projective kL-module and X is a relatively $\mathfrak{x}$-projective kG-module.

Let us see that this does generalize the theorem of the previous section. Suppose that Q is a trivial intersection subgroup so $gQg^{-1} \cap Q = 1$ if $g \notin N(Q)$. Hence, if $s \notin L$ then $sQs^{-1} \cap Q = 1$ so $\mathfrak{x} = \{1\}$ and $\mathfrak{z}$ consists of all non-identity subgroups of Q. Moreover, if Q is a Sylow p-subgroup then $sQs^{-1} \cap L = 1$ if $s^{-1} \notin L$: otherwise $sQs^{-1} \cap L$ is a p-subgroup of L so there is $x \in L$ with $x(sQs^{-1} \cap L)x^{-1} \subseteq Q$ so $xsQs^{-1}x^{-1} \cap Q \neq 1$ and $xs \notin L$, a contradiction. Thus, $\mathfrak{y} = \{1\}$ and both X and Y in the theorem are projective.

Let us also see that the expressions for U_L and V^G do determine the correspondence, just as stated above. In fact, we claim that no indecomposable summand of X or Y has vertex R. However, $R \in \mathfrak{z}$ so $R \nsubseteq_G \mathfrak{x}$ so this claim is clear for X. On the other hand, if there is an indecomposable summand of Y with vertex R then $R \subseteq_L \mathfrak{y}$, as Y is relatively $\mathfrak{y}$-projective. We need to deduce that $R \subseteq_G \mathfrak{x}$ and so derive a contradiction. But this is a consequence of the following result:

Lemma 2 *If R is a subgroup of Q then the following assertions are equivalent:*

   (1) $R \subseteq_G \mathfrak{x}$;
   (2) $R \subseteq_L \mathfrak{x}$;
   (3) $R \subseteq_L \mathfrak{y}$.

Proof If (1) holds then there is $g \in G$ with $gRg^{-1} \subseteq Q \cap sQs^{-1}$, where $s \in G$, $s \notin L$. If $g \in L$ then certainly (2) holds, while if $g \notin L$ then $R \subseteq g^{-1}Qg$ yields $R \subseteq Q \cap g^{-1}Qg$, that is, R is in $\mathfrak{x}$ so certainly $R \subseteq_L \mathfrak{x}$ and again (2) holds.

If (2) is valid then there is $x \in L$, $s \in G$, $s \notin L$ such that $xRx^{-1} \subseteq Q \cap sQs^{-1}$. Thus, trivially we have $xRx^{-1} \subseteq L \cap sQs^{-1}$ and $R \subseteq_L \mathfrak{y}$, that is, (3) is valid.

Finally, suppose that (3) holds. There is $x \in L$, $s \in G$, $s \notin L$ with $xRx^{-1} \subseteq L \cap sQs^{-1}$. Thus, $R \subseteq L \cap (x^{-1}s)Q(s^{-1}x)$. But $x^{-1}s \notin L$, so we have $R \subseteq_G x$, as $R \subseteq Q \cap (x^{-1}s)Q(s^{-1}x)$, and (1) holds.

The statement of the theorem is quite symmetrical in G and L, restriction and induction, but x and η are different. The next result gives the reason.

Lemma 3 *If U is a relatively x-projective kG-module then U_L is relatively η-projective. If V is a relatively η-projective kL-module then V^G is relatively x-projective.*

Proof Let W be an indecomposable summand of U so W is relatively projective for a subgroup of the form $sQs^{-1} \cap Q$, $s \notin L$. Hence, by Mackey's theorem, W_L is relatively projective for the collection of subgroups of the form

$$t(sQs^{-1} \cap Q)t^{-1} \cap L = tsQs^{-1}t^{-1} \cap tQt^{-1} \cap L.$$

But either $t \notin L$ or $ts \notin L$ (as $t \in L$, $s \notin L$ forces $ts \notin L$) so such a subgroup is contained in an element of η and so W_L and U_L are relatively η-projective.

On the other hand, if W is an indecomposable summand of V and W has vertex P then $P \subseteq_L \eta$ so W^G is relatively P-projective and $P \subseteq_G x$, by Lemma 2, so W^G is relatively x-projective and the lemma is proven.

We now turn to the proof of the theorem, the bulk of which is contained in the next two results.

Lemma 4 *If U is an indecomposable kG-module with vertex R in $\mathfrak{z}$ then $U_L \cong V \oplus Y$, where V is an indecomposable kL-module with vertex R, $U \mid V^G$ and Y is a relatively η-projective kL-module.*

Proof By Lemma 9.5, there is an indecomposable kL-module V with vertex R and $U \mid V^G$. Now $(V^G)_L \cong V \oplus Y_1$, where Y_1 is relatively η-projective, by Lemma 9.6, so U_L is either isomorphic with $V \oplus Y$ or Y for some summand Y of Y_1. However, again by Lemma 9.5, U_L has an indecomposable summand W with vertex R. Now W cannot be isomorphic with a summand of Y, or else $R \subseteq_L \eta$ and $R \subseteq_G x$, by Lemma 2, so $R \notin \mathfrak{z}$. Hence, $W \cong V$ and $U_L \cong V \oplus Y$, just as claimed.

Lemma 5 *If V is an indecomposable kL-module with vertex R in $\mathfrak{z}$ then $V^G \cong U \oplus X$ where U is an indecomposable kG-module with vertex R, $V \mid U_L$ and X is a relatively x-projective kG-module.*

Proof Let $V^G = U_1 + \cdots + U_r$ be a direct sum of indecomposable kG-modules. Since $(V^G)_L \cong V \oplus Y$, where Y is a relatively η-projective kL-module, by Lemma 9.6, we have, after renumbering, that $(U_1)_L \cong V + Y_1$, $(U_i)_L \cong Y_i$, $2 \leqslant i \leqslant r$, where the Y_i are kL-modules and $Y \cong Y_1 \oplus \cdots \oplus Y_r$. We claim that U_1 has a vertex in $\mathfrak{z}$ and that $U_2, \ldots, U_r$ are relatively $\mathfrak{x}$-projective. Indeed, U_1 has a vertex in Q, as $U_1 \mid V^G$, and U_1 cannot be relatively $\mathfrak{x}$-projective, as then Lemma 3 would imply that $(U_1)_L \cong V \oplus Y_1$ is relatively η-projective which is not the case. Hence, U_1 does have a vertex in $\mathfrak{z}$. Moreover, if U_i, $i > 1$, was not relatively $\mathfrak{x}$-projective, then Lemma 4 would apply to it and $(U_i)_L$ would not be relatively η-projective, which it is. Hence, U_i is indeed relatively $\mathfrak{x}$-projective.

Setting $U = U_1$, $X = U_2 + \cdots + U_r$ we have that $U \mid V^G$, $V^G \cong U \oplus X$ where X is relatively $\mathfrak{x}$-projective. It remains only to prove that U has vertex R. However, the vertex of U is in $\mathfrak{z}$ so Lemma 4 applies and there is a unique summand, in any decomposition of U_L into the direct sum of indecomposable modules, which is not relatively η-projective and that module has a vertex equal to a vertex of U. But $U_L \cong V \oplus Y$ so U has vertex R as V has vertex R. This proves the lemma.

It remains now only to prove Theorem 1. Everything has been established in the last two lemmas except the one-to-one property. We need to show two things: if U is an indecomposable kG-module with vertex R in $\mathfrak{z}$, V is as in Lemma 4, $V^G = U' \oplus Y$ as in Lemma 5, then $U \cong U'$; if we start with V a similar result holds. However, in Lemma 4 we proved that $U \mid V^G$ and in Lemma 5 that $V \mid U_L$, so everything is established.

Exercises

1   If $R \in \mathfrak{z}$ then $N_G(R) \subseteq L$.

2   If $R \in \mathfrak{z}$ then set
$$\mathfrak{x}_R = \{R \cap sRs^{-1} \mid s \in G, s \notin L\},$$
$$\eta_R = \{L \cap sRs^{-1} \mid s \in G, s \notin L\}.$$
State and prove an improvement of Theorem 1 using $\mathfrak{x}_R$ and η_R in place of $\mathfrak{x}$ and η.

3   If M is any kG-module then show there exist projective modules P_1 and P_2 and also modules L_1 and L_2, each a direct sum of module induced from p-local subgroups of G, such that
$$M \oplus P_1 \oplus L_1 \cong P_2 \oplus L_2.$$

12 Maps

We shall continue our study of the Green correspondence as a generalization of the trivial intersection case by examining homomorphisms between modules. A technical result on maps will lead us to a sort of converse to the Green correspondence and we shall conclude the section with the full extension of the final result of section 10. We shall preserve the notation of the previous section, in particular L, Q, x, η and $\mathfrak{z}$, and use it throughout this section.

A little more notation will be very useful for us. Suppose that $\mathscr{H}$ is a collection of subgroups of G. We shall denote the set of all kG-homomorphisms of the kG-module M to the kG-module N, which factors through relatively $\mathscr{H}$-projective module by

$$\mathrm{Hom}_{kG,\mathscr{H}}(M, N),$$

and any such map will be called $\mathscr{H}$-*projective*. The case that $\mathscr{H}$ consists of just the identity subgroup has already been considered (relatively $\{1\}$-projective is projective) and, just as we had then, we have that $\mathrm{Hom}_{kG,\mathscr{H}}(M, N)$ is a vector space. The quotient of $\mathrm{Hom}_{kG}(M, N)$ by this subspace is denoted by

$$\mathrm{Hom}_{kG}^{\mathscr{H}}(M, N).$$

The following technical result will give us our first theorem.

Lemma 1 *If M and N are kG-modules then*

$$\mathrm{Hom}_{kL,Q}(M, N) \subseteq \mathrm{Hom}_{kL,\eta}(M, N) + \mathrm{Hom}_{kG,Q}(M, N).$$

This is a statement about certain subspaces of $\mathrm{Hom}_{kL}(M, N)$ and shows that a Q-projective kL-homomorphism of M to N can be 'approximated' by a Q-projective kG-homomorphism with an 'error term' in $\mathrm{Hom}_{kL,\eta}(M, N)$.

Proof Let $\varphi \in \mathrm{Hom}_{kL,Q}(M, N)$ so there is a relatively Q-projective kL-module W and homomorphisms α and β such that the following diagram commutes:

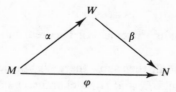

If i is the injection of W into W^G, sending $w \in W$ to $1 \otimes w$, then the definition of relatively free modules gives us another commutative diagram:

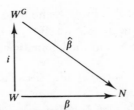

If π is the projection of W^G onto W, sending $1 \otimes w$ to w and $s \otimes w$ to zero if $s \notin L$, then, by exercise 8.3, we have another commuting diagram:

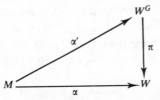

Putting these three diagrams together we get a bigger one (which is *not* commutative throughout):

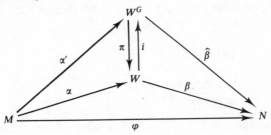

Hence, we have

$$\varphi = \beta\alpha$$
$$= \hat{\beta}\alpha' + (\beta\alpha - \hat{\beta}\alpha')$$
$$= \hat{\beta}\alpha' + \hat{\beta}i\pi\alpha' - \hat{\beta}\alpha'$$
$$= \hat{\beta}\alpha' + \hat{\beta}(i\pi - I)\alpha',$$

where I is the identity map of W^G. But $\hat{\beta}\alpha' \in \operatorname{Hom}_{kG, Q}(M, N)$ as it factors through W^G which is relatively Q-projective because W is. Thus, we need only show that

$$\hat{\beta}(i\pi - I)\alpha' \in \operatorname{Hom}_{kL, \mathfrak{y}}(M, N),$$

so that it suffices to see that $i\pi - I$ factors through a relatively $\mathfrak{y}$-projective kL-module. But clearly, by inspection, $i\pi - I$ factors through

$$U = \bigoplus_{\substack{s \in G/L \\ s \notin L}} s \otimes W$$

which is a kL summand of W^G. From Lemma 9.6, we have $(W^G)_L \cong W \oplus Y$, where Y is relatively η-projective. Since $(W^G)_L \cong W + U$, the Krull–Schmidt theorem implies that $U \cong Y$ and the lemma is proved.

Theorem 2 *Let U be an indecomposable kG-module with vertex Q and V the corresponding kL-module.*
 (1) *If M is a kG-module then $U \mid M$ if, and only if, $V \mid M_L$.*
 (2) *If M is an indecomposable kG-module and $V \mid M_L$ then $M \cong U$.*

The second statement is a sort of converse to the Green correspondence, in that it says if the restriction of M is 'right' then M is 'right', rather than the reverse of this. This statement also implies the first statement (and is also trivially a consequence of it) since we can apply (2) to each indecomposable summand of M.

  This result also tells us about the 'error term' X, where $V^G \cong U \oplus X$ is the Green correspondence. Indeed, suppose that U_1 is an indecomposable kG-module with vertex Q_1, $L_1 = N_G(Q_1)$ and V_1 is a kL_1-module which is a Green correspondence of U_1. Thus, $U_1 \mid V^G$ if, and only if, $V_1 \mid (V^G)_{L_1}$. Hence, if $U_1 \not\cong U$, then $U \mid X$ if, and only if,

$$V_1 \ \Big| \bigoplus_{s \in L_1 \backslash G/L} (s(V)_{sLs^{-1} \cap L_1})^{L_1},$$

a question involving subgroups of G.
  It remains now to prove (2).

Proof We may assume that $M_L = V + W$ is a direct sum so the projection π of M_L onto V with kernel W is in $\mathrm{Hom}_{kL,Q}(M, M)$ as it factors through V. Express $\pi = \alpha + \beta$, using Lemma 2, where $\alpha \in \mathrm{Hom}_{kG,Q}(M, M)$ and $\beta \in \mathrm{Hom}_{kL,\eta}(M, M)$.

  Suppose next that $\mathrm{Hom}_{kG,Q}(M, M)$ is properly contained in $\mathrm{Hom}_{kG}(M, M) = \mathrm{End}(M)$. It is clearly an ideal, as a homomorphism that factors through a relatively Q-projective module when composed with any homomorphism results in a homomorphism with the same property. But $\mathrm{End}(M)$ is a local algebra so this proper ideal is contained in the radical. We deduce that $\alpha^n = 0$ for some positive integer n. Now $\pi = \pi^n$ and $\pi^n - (\alpha + \beta)^n$ is a sum of 2^n terms, one being $\alpha^n = 0$, the rest all involving β, so $\pi \in \mathrm{Hom}_{kL,\eta}(M, M)$ as $\beta \in \mathrm{Hom}_{kL,\eta}(M, M)$. This implies that the identity map I_V of V lies in $\mathrm{Hom}_{kL,\eta}(V, V)$: for I_V is the composition of an imbedding of V in M, π (which is an endomorphism of M) and a homomorphism onto V, while $\pi \in \mathrm{Hom}_{kL,\eta}(M, M)$. Therefore, there is a relatively η-projective kL-module Y such that I_V factors through Y. That is, there is $\lambda \in \mathrm{Hom}_{kL}(V, Y)$,

$\mu \in \text{Hom}_{kL}(Y, V)$ with $\mu\lambda = I_V$, so $V \mid Y$. But V has vertex Q and $Q \notin \eta$, so we have a contradiction. Therefore, we deduce that $\text{End}(M) = \text{Hom}_{kG, Q}(M, M)$.

In particular, the identity map I_M of M is Q-projective so M is a direct summand of a relatively Q-projective module, that is, M is relatively Q-projective. Moreover, Q is the vertex of M, since otherwise the vertex of M would have order strictly smaller than $|Q|$ and, by Mackey's theorem, each indecomposable summand of M_L would be relatively projective for a subgroup of order less than $|Q|$. This is not the case as $V \mid M_L$. We can now apply the Green correspondence to M. We deduce that M_L has a unique indecomposable summand with vertex Q and it is the correspondent of M. But $V \mid M_L$ so M corresponds to V, that is, $M \cong U$.

The next result is the definitive one on maps.

Theorem 3 *If U is an indecomposable kG-module with vertex R in $\mathfrak{z}$, V is the corresponding kL-module and M is any kG-module then*

$$\text{Hom}^{\mathfrak{x}}_{kG}(M, U) \cong \text{Hom}^{\mathfrak{x}}_{kL}(M_L, V).$$

This looks a little surprising: one expects η and not $\mathfrak{x}$ when dealing with L. But, in fact, $\text{Hom}_{kL, \mathfrak{x}}(M_L, V) = \text{Hom}_{kL, \eta}(M_L, V)$. One other point should be made now: we will be able to deduce, as a consequence, a result which generalizes the last result of section 10.

Another useful technical result is next.

Lemma 4 *If U, V and W are kG-modules, $\mathcal{H}$ and $\mathcal{K}$ are collections of subgroups of G, α is an $\mathcal{H}$-projective homomorphism of U to V and β is a $\mathcal{K}$-projective homomorphism of V to W then $\beta\alpha$ is an $\mathcal{L}$-projective homomorphism of U to W where $\mathcal{L}$ consists of all subgroups of the form $H \cap sKs^{-1}$, $H \in \mathcal{H}$, $K \in \mathcal{K}$, $s \in G$.*

Proof Since every relatively $\mathcal{H}$-projective module is the direct sum of modules each of which is relatively H-projective for some $H \in \mathcal{H}$, it follows that each $\mathcal{H}$-projective homomorphism of U to V is the sum of maps, each of which factors through a relatively H-projective module for some $H \in \mathcal{H}$. But here we can use modules induced from H as every relatively H-projective module is a summand of such an induced module. Since the same holds for $\mathcal{K}$ we need only consider the following situation,

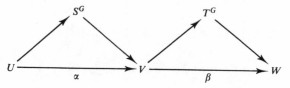

where the triangles are commutative, S is a kH-module, $H \in \mathcal{H}$, T is a kK-module, $K \in \mathcal{K}$, and prove that $\beta\alpha$ is projective for the set of subgroups $H \cap sKs^{-1}$, $s \in G$.

However, consider the map φ which is the kG-homomorphism of S^G to T^G which is the composition of the given maps of S^G to V and V to T^G. It suffices to show that φ is projective for the set of intersections just described. However, the kH-module $(T^G)_H$ is relatively projective for this set of intersections, by Mackey's theorem, so the kG-module $((T^G)_H)^G$ is also relatively projective for this collection. Hence, it suffices to show that φ factors through $((T^G)_H)^G$. We do this as a separate result:

Sublemma *Let H be a subgroup of G, W a kH-module and V a kG-module. If $\varphi \in \mathrm{Hom}_{kG}(W^G, V)$ then φ factors through $(V_H)^G$.*

Proof Let i be the injection of W into W^G sending $w \in W$ to $1 \otimes w$ and let π be the usual projection of W^G onto W sending $1 \otimes w$ to w, $s \otimes w$ to zero if $s \notin H$. Set $\psi = \varphi i \pi$ so $\psi \in \mathrm{Hom}_{kH}(W^G, V)$. Let $\psi' \in \mathrm{Hom}_{kG}(W^G, (V_H)^G)$ be defined as in part (3) of Lemma 8.6: if $u \in W^G$ then $\psi'(u) = \sum_{s \in G/H} s \otimes \psi(s^{-1}u)$. Let ρ be the kG-homomorphism of $(V_H)^G$ onto V such that $\rho(\sum_{t \in G/H} t \otimes v_t) = \sum tv_t$: this is easily seen to be a kG-homomorphism. We now claim that the following diagram commutes:

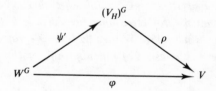

This assertion will, of course, complete the proof. It needs only a direct calculation. Let $\sum_{t \in G/H} t \otimes w_t \in W^G$, each $w_t \in W$. Thus

$$\varphi\left(\sum t \otimes w_t\right) = \sum (t(1 \otimes w_t)) = \sum t\varphi(1 \otimes w_t).$$

Moreover,

$$\rho\psi'\left(\sum t \otimes w_t\right) = \rho\left(\sum_{s \in G/H} s \otimes \psi\left(s^{-1}\sum_t t \otimes w_t\right)\right)$$

$$= \rho\left(\sum_s s \otimes \varphi i \pi\left(s^{-1}\sum_t t \otimes w_t\right)\right)$$

$$= \rho\left(\sum_s s \otimes \varphi i(w_s) \right)$$

$$= \rho\left(\sum_s s \otimes \varphi(1 \otimes w_s) \right)$$

$$= \sum_s s\varphi(1 \otimes w_s),$$

as required.

Let us now turn to the proof of Theorem 3. Every $\mathfrak{y}$-projective homomorphism of M_L to V equals its composition with the identity maps of V, which is Q-projective since V is certainly relatively Q-projective, so

$$\mathrm{Hom}_{kL, \mathfrak{y}}(M_L, V) = \mathrm{Hom}_{kL, \mathfrak{x}}(M_L, V)$$

by the previous lemma, as the intersection of a subgroup of $\mathfrak{y}$ with Q is a subgroup of $\mathfrak{x}$ and each subgroup in $\mathfrak{x}$ so arises. Hence, it suffices to prove that

$$\mathrm{Hom}_{kG}^{\mathfrak{x}}(M, U) \cong \mathrm{Hom}_{kL}^{\mathfrak{y}}(M_L, V).$$

But $V^G \cong U \oplus X$, where X is relatively $\mathfrak{x}$-projective, so

$$\mathrm{Hom}_{kG}^{\mathfrak{x}}(M, U) \cong \mathrm{Hom}_{kG}^{\mathfrak{x}}(M, V^G).$$

Hence, we need only establish an isomorphism

$$\mathrm{Hom}_{kG}^{\mathfrak{x}}(M, V^G) \cong \mathrm{Hom}_{kL}^{\mathfrak{y}}(M_L, V).$$

We certainly have that

$$\mathrm{Hom}_{kG}(M, V^G) \cong \mathrm{Hom}_{kL}(M_L, V)$$

so it is enough that the $\mathfrak{x}$-projective homomorphism on the left-hand side corresponds to the $\mathfrak{y}$-projective homomorphisms on the right-hand side under the isomorphism.

First, suppose that $\alpha \in \mathrm{Hom}_{kL, \mathfrak{y}}(M_L, V)$ so $\alpha \in \mathrm{Hom}_{kL, \mathfrak{x}}(M_L, V)$. Thus, there is a relatively $\mathfrak{x}$-projective kL-module W and homomorphisms β, γ such that the following diagram commutes:

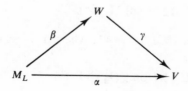

Hence, by the Omnibus Lemma, we have a commutative diagram

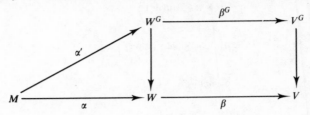

so the corresponding homomorphism of M to V^G factors through the relatively x-projective module V^G.

Second, suppose $\alpha \in \mathrm{Hom}_{kL}(M, V)$ and that the corresponding homomorphism $\alpha' \in \mathrm{Hom}_{kG}(M, V^G)$ factors through a relatively projective kG-module N, so we have a commutative diagram as follows:

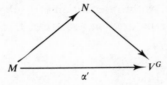

Now N_L is relatively η-projective, by Lemma 11.3, so α', as a kL-homomorphism, is η-projective. But we have a commutative diagram (see exercise 8.3)

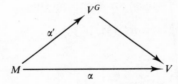

so putting these last two diagrams together we get that α factors through the η-projective module N_L. Hence, the theorem is proved.

Finally, we can generalize the last result of section 10.

Corollary 5 *If U_1 and U_2 are indecomposable kG-modules with vertices in $\mathfrak{z}$ and V_1 and V_2 are the corresponding kL-modules then*

$$\mathrm{Hom}_{kG}^{x}(U_1, U_2) \cong \mathrm{Hom}_{kL}^{x}(V_1, V_2).$$

Proof We know that $(U_1)_L \cong V_1 \oplus Y$, where Y is relatively η-projective. Hence, by Theorem 3,

$$\mathrm{Hom}_{kG}^{x}(U_1, U_2) \cong \mathrm{Hom}_{kL}^{x}(V_1 \oplus Y, V_2)$$

But V_2 is relatively Q-projective so, again by Lemma 4, any kL-homomorphism of Y to V_2 is $\mathfrak{x}$-projective. Thus,

$$\mathrm{Hom}^{\mathfrak{x}}_{kL}(V_1 + Y, V_2) \cong \mathrm{Hom}^{\mathfrak{x}}_{kL}(V_1, V_2)$$

and we are done.

IV

Blocks

Up until now we have considered the totality of all kG-modules and studied their structure directly or through the use of local subgroups. However, by examining direct decompositions of the algebra kG, we shall introduce the notion of a block and this will enable us to partition the set of isomorphism classes of indecomposable modules in a very useful way. Again local subgroups play an important role and we can relate the new ideas with those in the previous chapter. This will lead us up to the final chapter and the deepest results of the whole theory.

13 Defect groups

We shall introduce decompositions of the algebra A into the direct sum of algebras, prove an easy but vital uniqueness statement and see how this is reflected in modules. We shall then specialize to group algebras and introduce certain relevant p-subgroups which play a central role.

Theorem 1 *The algebra A has a unique decomposition into a direct sum of subalgebras each of which is indecomposable as an algebra.*

If the unique decomposition of A is given by

$$A = A_1 + \cdots + A_r$$

then the subalgebras $A_1, \ldots, A_r$ are called the *blocks* of A. Each A_i is an ideal and multiplication is componentwise: if $a_i \in A_i$, $a_j \in A_j$, $i \neq j$ then $a_i a_j = 0$ as $a_i a_j \in A_i \cap A_j$.

Proof Express $A = A_1 + \cdots + A_r$ as a direct sum of subalgebras, each of which is indecomposable as an algebra. Let B be such a summand in another such decomposition; it suffices to prove that $B = A_i$, for some i, $1 \leqslant i \leqslant r$. Let $b \in B$ and express $b = a_1 + \cdots + a_r$, each $a_i \in A_i$. Since A_i has a unit element, the component of the unit element 1 of A in A_i, and B is an ideal, it follows that $a_i \in B$ for each i. Hence,

$$B = (B \cap A_1) + \cdots + (B \cap A_r)$$

and this is a decomposition of B, inasmuch as each $B \cap A_i$ is an ideal of B. By our assumption on B we must have that all these summands, in the expression for B, are zero, with one exception. Hence, $B \subseteq A_i$ for some i. However, by the same argument, applied to A_i in comparison with the decomposition of A involving B, we must have $A_i \subseteq B$ and the theorem is proven.

We shall preserve the notation $A = A_1 + \cdots + A_r$ for the decomposition of A into blocks. If M is an A-module, $A_i M = M$ and $A_j M = 0$ for all j, $j \neq i$, then we say that M *lies in the block* A_i. Conversely, suppose that N is an A_i-module; we can make N into an A-module, by simply demanding that each A_j, $j \neq i$, annihilates N, and it lies in A_i. Each A-module lying in A_i clearly arises in this way.

There is another way to express this. Let $1 = e_1 + \cdots + e_r$ be the components of the unit element 1 so e_j is the identity element of A_j. In particular, since 1 induces the identity on each A-module, if M lies in A_i, then e_i is the identity on M and each e_j, $j \neq i$, annihilates M. On the other hand, if this occurs then M is in A_i, inasmuch as $AM = Ae_i M = A_i M$ and $A_j M = A_j e_j M$. This criterion for lying in a block immediately implies that submodules, quotient modules and direct sums of modules lying in a block A_i also lie in A_i. Moreover, if M_i and M_j lie in the blocks A_i and A_j, respectively, and $i \neq j$ then $\operatorname{Hom}_A(M_i, M_j) = 0$. Indeed, if $\varphi \in \operatorname{Hom}_A(M_i, M_j)$ then e_i is the identity on $\varphi(M_i)$ but $e_i M_j = 0$.

The next result shows the importance of the idea of lying in a block.

Proposition 2 *If M is an A-module then M has a unique direct decomposition*

$$M = M_1 + \cdots + M_r,$$

where M_i lies in the block A_i.

Proof We set $M_i = A_i M$ so certainly M_i is an A-module and lies in the block A_i while M is the sum of $M_1, \ldots, M_r$ since $1 \cdot M = M$. Moreover, the sum is direct. Indeed, suppose that $m_1 + \cdots + m_r = 0$, where $m_i \in M_i$. Applying the unit element of A_j to this sum yields $m_j = 0$, for each j, $1 \leqslant j \leqslant r$.

On the other hand, suppose $M = N_1 + \cdots + N_r$ is a direct sum where N_i

lies in A_i. Thus, $N_i = A_i N_i \subseteq A_i M = M_i$ so $N_i \subseteq M_i$ and hence $N_i = M_i$, since M is the sum of $N_1, \ldots, N_r$.

We now know in a precise way how the study of A-modules reduces to the study of A_i-modules, $1 \leqslant i \leqslant r$. In due course, we shall also see how, for the case of group algebras, a block can be analyzed using local subgroups and how powerful this reduction is.

The next problem we wish to deal with is how to determine the blocks in a useful way, and we shall do this in terms of simple modules. If M is an A-module lying in the block A_i then each composition factor, being a quotient of a submodule of M, also lies in A_i. Conversely, if each composition factor of M lies in A_i then the unit element of A_i induces the identity on each such factor and the unit element of A_j, $j \neq i$, annihilates each factor and so the same can be said for M and so M lies in A_i. Hence, to determine the blocks via modules, it suffices to know when two simple modules lie in the same block.

Proposition 3 *If S and T are simple A-modules then the following are equivalent:*

    (1) *S and T lie in the same block;*

    (2) *There are simple A-modules*

$$S = S_1, S_2, \ldots, S_m = T$$

    such that S_i, S_{i+1}, $1 \leqslant i < m$, are composition factors of an indecomposable projective A-module;

    (3) *There are simple A-modules*

$$S = T_1, T_2, \ldots, T_n = T$$

    such that T_i, T_{i+1}, $1 \leqslant i < n$, are equal or there is a non-split extension of one of them by the other.

The second condition applies immediately to kG, $G = \mathrm{SL}(2, p)$. Recall that $V_1, \ldots, V_p$ are the simple kG-modules, V_p is projective and that each indecomposable projection module has all its composition factors of even dimension or all of odd dimension. In fact, as is easy to show (see the exercises), the simple modules fall into blocks as follows: $V_1, V_3, \ldots, V_{p-2}$; $V_2, \ldots, V_{p-1}$; V_p (except for $p = 2$).

The second condition also tells us how to determine the decomposition of A into blocks if we have a decomposition of A into the direct sum of indecomposable projective modules. Indeed, suppose

$$A = P_1 + \cdots + P_k$$

is such a decomposition. Suppose that $P_1, \ldots, P_j$ all lie in a block and none of the other summands lie in that block. Then the block equals $P_1 + \cdots + P_j$. In fact, if B is the block in question then $B = BA$ while $BP_i = P_i$, $1 \leqslant i \leqslant j$, while $BP_i = 0$ if $i > j$.

Proof The three conditions describe three relations on the set of simple A-modules and we must see that these are all the same relation. Now if T and T' are simple A-modules and there is a non-split extension M of T' by T' then M is a uniserial module so it is a homomorphic image of the indecomposable projective module corresponding to T. Thus, the relation given by (3) implies the relation given by (2). Moreover, by Proposition 2, any indecomposable module lies in a block, so the relation given by (2) implies the one given by (1). We must now reason in the reverse order.

We shall first show that the relation for (1) implies the second relation. Let S and T be simple A-modules which lie in the block B of A and assume that they are not as in (2). In particular, we can express

$$B = P_1 + \cdots + P_s + Q_1 + \cdots + Q_t$$

as a direct sum of indecomposable projective modules, where each composition factor of P_i is related to S by (2) and no composition factor of any Q_j is related to S by (2). Indeed, this holds because S and T are homomorphic images of indecomposable projective B-modules as they are simple B-modules. This implies that $\operatorname{Hom}_B(P_i, Q_j) = 0$ and $\operatorname{Hom}_B(Q_j, P_i) = 0$, since no composition factor of any Q_j can be a composition factor of any P_i. Thus, $P_1 + \cdots + P_s$ and $Q_1 + \cdots + Q_t$ are invariant under all endomorphisms of B, that is, these sums are ideals, as the endomorphisms are given by right multiplications. This is a contradiction.

Finally, let us show that the relation given by (2) implies the one for (3). It suffices to show that if V is a simple A-module and P is the corresponding indecomposable projective module then V is related to each of the composition factors of P by the relation given by (3). Indeed, if S and T are as in (2) then one can easily expand the sequence given, by adding, between S_i and S_{i+1}, the simple module corresponding to an indecomposable projective module which has S_i and S_{i+1} as common composition factors, and still have a sequence satisfying (2). Now, if we prove our claim, it is clear that S and T are related as in (3).

Therefore, in order to finish the proof it is sufficient to verify the following statement: if W is a composition factor of $\operatorname{rad}^{i+1}(P)/\operatorname{rad}^{i+2}(P)$ then there is a composition factor U of $\operatorname{rad}^i(P)/\operatorname{rad}^{i+1}(P)$ and a non-split extension of W by U. Hence, we need only prove the following, more general, assertion and apply it to $\operatorname{rad}^i(P)/\operatorname{rad}^{i+2}(P)$: if X is an A-module and $\operatorname{rad}(X)$ is semisimple then for any composition factor W of $\mathrm{rad}(X)$ there is a composition factor U of $X/\mathrm{rad}(X)$ and a non-split extension of W by U. We may assume that $\mathrm{rad}(X) = W + W'$, a direct sum, as $\mathrm{rad}(X)$ is semisimple. The module X/W' satisfies the same conditions, so, without loss of generality, we may assume $\mathrm{rad}(X) = W$. There are submodules $Y_1, \ldots, Y_n$ of X containing W such that X/W is the direct sum of the simple modules Y_i/W. Now $\mathrm{rad}(X) = (\mathrm{rad}\ A)X$ so $\mathrm{rad}(X)$ is the sum of all $(\mathrm{rad}\ A)Y_i = \mathrm{rad}(Y_i)$. Hence, there is j, $1 \leqslant j \leqslant n$, such that $\mathrm{rad}(Y_j) \neq 0$. Since $W = \mathrm{rad}(X)$ is simple, we have that $\mathrm{rad}(Y_j) = W$ and so Y_j is a non-split extension of W by the simple module Y_j/W.

We shall now specialize to the case of group algebras. The key idea is to regard kG as a module for the group algebra $k[G \times G]$: if $a \in kG$, $g_1, g_2 \in G$ then set $(g_1, g_2)a = g_1 a g_2^{-1}$. Therefore, the submodules of the $k[G \times G]$-module kG are exactly the ideals of kG. In particular, kG has a unique decomposition into the direct sum of indecomposable $k[G \times G]$-modules; the summands are the blocks of kG. These blocks, as $k[G \times G]$-modules, are pairwise non-isomorphic: their annihilators in $k[G \times 1] \subseteq k[G \times G]$ are different.

The idea now is to apply our knowledge of indecomposable modules in this situation. The homomorphism δ of G into $G \times G$, sending g in G to (g, g), will be very useful.

Theorem 4 *If B is a block of kG then B has a vertex, as a $k[G \times G]$-module, of the form δD, where D is a p-subgroup of G.*

If H and K are subgroups of G with δH and δK conjugate in $G \times G$ then H and K are conjugate in G: this is easy to see by projecting on the first coordinate in $G \times G$. Hence, the subgroups D of G, such that δD is a vertex of the $k[G \times G]$-module B, are a conjugacy class of p-subgroups of G. These are called the *defect groups* of B. The terminology is due to the fact that the size of D is a measure of the deviation of B, as an algebra, from being semisimple. If D has order p^d then B is said to be of *defect d*.

Proof It suffices to show that the $k[G \times G]$-module B is relatively δG-projective since then δG contains a vertex of B. But B is a direct summand of the $k[G \times G]$-module kG, so we need only demonstrate that kG is relatively δG-projective. But kG contains the subspace $k \cdot 1$, which is a $k[\delta G]$-module, in fact is a trivial $k[\delta G]$-module. Moreover,

$$\dim_k kG = |G| \dim_k(k \cdot 1) = |G \times G : \delta(G)| \dim_k(k \cdot 1)$$

and it is easy to see that $k \cdot 1$ generates the $k[G \times G]$-module kG. Hence, by our characterization of induced modules, $kG \cong (k \cdot 1)^{G \times G}$ so certainly kG is relatively $\delta(G)$-projective.

The defect group of a block is related to the modules in the block.

Theorem 5 *If B is a block of G and D is a defect group of B then any indecomposable kG-module lying in B has a vertex contained in D.*

In fact, this is part of a characterization of defect groups. As we shall see in the next section, D is a vertex of an indecomposable module in B, so D is characterized as being one of the largest vertices that occurs for indecomposable B-modules. Once this is established, we will know, in particular, that B is semisimple, as an algebra, if, and only if, B has defect 0. Indeed, B is semisimple if, and only if, each B-module is projective (so all exact sequences split), that is all vertices of indecomposable B-modules are the identity subgroup.

Proof We begin by considering B as a kG-module using conjugation: if $g \in G$ and $\beta \in B$ then $g \cdot b = g\beta g^{-1}$. If U is any kG-module then there is a linear transformation of the kG-module $B \otimes U$ (with the usual diagonal action) to the kG-module U which sends $\beta \otimes u$, for $\beta \in B$ and $u \in U$, to the product βu, because of the bilinearity of this formula. Moreover, this transformation is a kG-homomorphism: if $g \in G, \beta \in B, u \in U$ then $g(\beta \otimes u) = g\beta g^{-1} \otimes gu$, which is mapped to $g\beta g^{-1}gu = g(\beta u)$ and this is just g applied to the image of $\beta \otimes u$.

There is a kG-homomorphism of U to $B \otimes U$. Let e be the unit element of B so there is a linear transformation of U to $B \otimes U$ which sends $u \in U$ to $e \otimes u$. This is a kG-homomorphism, inasmuch as gu, for $g \in G$, is sent to $e \otimes gu$ and $e \otimes gu = g(e \otimes u)$ as $g^{-1}eg = e$, e being in the center of kG. If U lies in the block B then $eu = u$, for each $u \in U$ so the composition of these two kG-homomorphisms sends u to $e \otimes u$ to $eu = u$. Hence, $U \mid B \otimes U$ and so it suffices to prove that $B \otimes U$ is relatively D-projective as then U is also. However, by Lemma 8.5, part (5) (see exercise 9.4), it suffices to prove that the kG-module B is relatively D-projective as then the tensor product $B \otimes U$ is as well.

The isomorphism $G \cong \delta(G)$ allows us, by transport of structure, to consider B as a $k[\delta(G)]$-module: $(g, g)\beta = g\beta g^{-1}$, for $g \in G$, $\beta \in B$. Hence, we need only prove that B as a $k[\delta(G)]$-module is relatively $\delta(D)$-projective. However, we are just looking at the $k[G \times G]$-module B restricted to $\delta(G)$. We know that B, as a $k[G \times G]$-module, is induced from a $k[\delta(G)]$-

submodule, so, by Mackey's theorem, B, as a $k[\delta(G)]$-module, is relatively projective for the collection of subgroups

$$\delta(G) \cap (g_1, g_2) \, \delta(D)(g_1, g_2)^{-1}$$

where g_1 and g_2 run over G. It suffices therefore to prove that each of these subgroups is conjugate in $\delta(G)$ with a subgroup of $\delta(D)$. Moreover, if $d \in D$, $g_1, g_2 \in G$ and $(g_1, g_2) \, \delta(d)(g_1, g_2)^{-1} \in \delta(G)$, that is, $g_1 dg_1^{-1} = g_2 dg_2^{-1}$ and so

$$
\begin{aligned}
(g_1, g_2) \, \delta(d)(g_1, g_2)^{-1} &= (g_1 dg_1^{-1}, g_2 dg_2^{-1}) \\
&= (g_1 dg_1^{-1}, g_1 dg_1^{-1}) \\
&= \delta(g_1) \, \delta(d) \, \delta(g_1)^{-1}
\end{aligned}
$$

is contained in $\delta(g_1) \, \delta(D) \, \delta(g_1)^{-1}$, a conjugate of $\delta(D)$ in $\delta(G)$, just as required.

Our final result shows how defect groups are far from arbitrary.

Theorem 6 *If B is a block of G and D is a defect group of B then the following hold:*

(1) *If S is a Sylow p-subgroup of G containing D then there is a $c \in C(D)$ such that $D = S \cap cSc^{-1}$;*

(2) *D contains every normal p-subgroup of G;*

(3) *D is the largest normal p-subgroup of $N_G(D)$.*

The first statement implies the other two. Indeed, suppose that (1) holds and let N be a normal p-subgroup of G. By Sylow's theorem, N is contained in every Sylow p-subgroup of G, so $N \subseteq S$ and $N \subseteq cSc^{-1}$. Hence, $N \subseteq D$. Next, let T be a Sylow p-subgroup of $N_G(D)$ and choose S containing T. Thus,

$$D \subseteq T \cap cTc^{-1} \subseteq S \cap cSc^{-1} = D.$$

But cTc^{-1} is also a Sylow p-subgroup of $N_G(D)$ so D is the intersection of Sylow p-subgroups of $N_G(D)$. Since every normal p-subgroup of $N_G(D)$ is contained in every Sylow p-subgroup of $N_G(D)$, we have established the third statement.

In order to prove (1), we shall study kG as a $k[S \times S]$-module (that is, restrict from $G \times G$ to $S \times S$). We need some properties of such restrictions for this proof and later as well, so we shall gather the useful results together.

Lemma 7 *Let H be a subgroup of G and t an element of G.*

(1) *The $k[H \times H]$-submodule $kHtH$ of kG is induced from the trivial submodule of*

$$(1, t)^{-1} \, \delta(H \cap tHt^{-1})(1, t).$$

(2) *If H is a p-subgroup of G then kHtH is indecomposable as a $k[H \times H]$-module and its vertex is*

$$(1, t)^{-1} \delta(H \cap tHt^{-1})(1, t).$$

(3) *If Q is a p-subgroup of H, $C_G(Q) \subseteq H$ and $t \notin H$ then no indecomposable summand of kHtH has a vertex containing $\delta(Q)$.*

Here the module $kHtH$ consists of all linear combinations of the elements of the double coset HtH. To prove the lemma, we consider the subgroup of $H \times H$ consisting of the elements leaving t fixed. If $h_1, h_2 \in H$ then $(h_1, h_2)t = t$ if, and only if, $h_1 t h_2^{-1} = t$, that is, $h_2 = t^{-1} h_1 t$. Hence, this subgroup consists of all pairs $(h, t^{-1} ht)$ such that $h \in H$ and $t^{-1} ht \in H$, that is, $h \in H \cap tHt^{-1}$. Hence, the subgroup consists of all elements

$$(1, t)^{-1}(h, h)(1, t)$$

where $h \in H \cap tHt^{-1}$, that is, the subgroup is

$$(1, t)^{-1} \delta(H \cap tHt^{-1})(1, t).$$

The dimension of $kHtH$ is the product of $|H|$ and the number of cosets in the double coset HtH so

$$\dim kHtH = |H||H:H \cap tHt^{-1}|$$
$$= |H \times H:(1, t)^{-1} \delta(H \cap tHt^{-1})(1, t)|.$$

Moreover, it is clear that t generates $kHtH$ as a $k[H \times H]$-module so, by our characterization of induced modules, we have proved (1). The second statement is now immediate, as now $kHtH$ is the module induced from the trivial module of subgroups of a p-subgroup so is indecomposable with the subgroup as a vertex (see the discussion following the proof of Theorem 9.4 for an implicit proof).

Turning to (3), the first statement yields that each indecomposable summand of $kHtH$ has a vertex contained in $(1, t)^{-1} \delta(H \cap tHt^{-1})(1, t)$. Hence, if (3) is false then there is a conjugate, in $H \times H$, of $\delta(Q)$ which is contained in $(1, t)^{-1} \delta(H \cap tHt^{-1})(1, t)$. In particular, there will be h_1, h_2 in H such that

$$(h_1, h_2)^{-1} \delta(Q)(h_1, h_2) \subseteq (1, t)^{-1} \delta(G)(1, t),$$

that is,

$$(1, t)(h_1, h_2)^{-1} \delta(Q)(h_1, h_2)(1, t)^{-1} \subseteq \delta(G).$$

This means that if $x \in Q$, then

$$h_1^{-1} x h_1 = t h_2^{-1} x h_2 t$$

so $h_1 t h_2^{-1} \in C(Q)$ and $t \in h_1^{-1} C(Q) h_2 \subseteq H$, a contradiction.

Let us now complete the proof of the theorem. Since $\delta(D) \subseteq S \times S$ and B, as a $k[G \times G]$-module, has $\delta(D)$ as a vertex, it follows from Lemma 9.5 that the restriction of B to $S \times S$ has an indecomposable summand with vertex $\delta(D)$. But B is a summand of kG and

$$(kG)_{S \times S} \cong \bigoplus_{t \in S \backslash G / S} kStS,$$

since G is a disjoint union of all the S, S double cosets. But the lemma shows that each $k[S \times S]$-module $kStS$ is indecomposable and has $(1, t)^{-1} \delta(S \cap tSt^{-1})(1, t)$ as a vertex. Hence, there is $t \in G$ such that $\delta(D)$ is conjugate in $S \times S$ with $(1, t)^{-1} \delta(S \cap tSt^{-1})(1, t)$. That is, there exist $r, s \in S$ with

$$\delta(D) = (r, s)(1, t)^{-1} \delta(S \cap tSt^{-1})(1, t)(r, s)^{-1}.$$

In particular, $|D| = |S \cap tSt^{-1}|$. Moreover, this implies that

$$(1, t)(r, s)^{-1} \delta(D)(r, s)(1, t)^{-1} \subseteq \delta(G)$$

so, for any $d \in D$,

$$r^{-1}dr = ts^{-1}dst^{-1},$$

that is, $rts^{-1} \in C(D)$. Set $c = rts^{-1}$; it suffices to show that $D = S \cap cSc^{-1}$. Certainly D is contained in this intersection, as $D \subseteq S$ and $c \in C(D)$, so we need only show that D and $S \cap cSc^{-1}$ have the same order. However,

$$|S \cap cSc^{-1}| = |S \cap rts^{-1}Sst^{-1}r^{-1}|$$
$$= |r^{-1}Sr \cap tSt^{-1}|$$
$$= |S \cap tSt^{-1}|$$
$$= |D|.$$

Exercises

1   If S and T are simple A-modules then S and T lie in the same block if, and only if, there is a sequence

$$S = S_1, \ldots, S_n = T$$

of simple A-modules such that $\text{Hom}_A(P_i, P_{i+1}) \neq 0$ or $\text{Hom}_A(P_{i+1}, P_i) \neq 0, 1 \leq i < n$, where P_j is the indecomposable projective module corresponding to S_j, $1 \leq j \leq n$.

2   If $1 \leq i, j < p$ then the simple modules V_i, V_j for $SL(2, p)$ lie in the same block if, and only if, i and j are both even or are both odd.

3   Assume that R is a cyclic normal Sylow p-subgroup of G and use the notation of section 5. Show that the simple kG-modules S and T lie in

the same block if, and only if,

$$T \cong S \otimes W \otimes \cdots \otimes W$$

(for some number of factors W). (Hint: Show that $W \otimes \cdots \otimes W \cong k$ for a number of factors.)

14 Brauer correspondence

We shall now show how to determine the blocks locally and to relate this information to the Green correspondence. The key idea is a method of relating blocks of G with blocks of subgroups. In fact, suppose that H is a subgroup of G, b is a block of H and B is a block of G. We say that B *corresponds to* b and write $B = b^G$ (this is not induction) provided that b, as a $k[H \times H]$-module, is a summand of the restriction $B_{H \times H}$ from $G \times G$ *and* that B is the only block of G with this property. We have therefore defined a map from a subset of the blocks of H to the blocks of G. If b is a block of H such that it has a block of G corresponding to it, we also say that b^G is defined.

Let us first prove some basic facts about this correspondence.

Lemma 1 *Let b be a block of the subgroup H of G and let D be a defect group of b.*

(1) *If b^G is defined then D is contained in a defect group of b^G.*

(2) *If H is contained in the subgroup K of G while b^K, $(b^K)^G$ and b^G are defined, then $b^G = (b^K)^G$.*

(3) *If $C_G(D) \subseteq H$ then b^G is defined.*

Proof Let E be a defect group of B so the $k[G \times G]$-module B has vertex $\delta(E)$. Since $b \mid B_{H \times H}$ it follows from Mackey's theorem that the vertex $\delta(D)$ of b is conjugate in $G \times G$ with a subgroup of $\delta(E)$. However, if $(g_1, g_2) \in G \times G$ conjugates $\delta(D)$ into $\delta(E)$ then g_1 conjugates D into E so (1) is proven. The second assertion is immediate from the definition since $b \mid (b^G)_{H \times H}$ and $b \mid ((b^K)^G)_{H \times H}$, the latter since $b \mid (b^K)_{H \times H}$ and $b^K \mid ((b^K)^G)_{K \times K}$.

In order to establish (3), we need only see that the hypothesis implies that b occurs exactly once in a decomposition of $(kG)_{H \times H}$ into a direct sum of indecomposable modules. But kH is a direct sum of the blocks of H, no two of which are isomorphic as $k[H \times H]$-modules; hence, it is sufficient to show that $b \mid kHtH$ when $t \notin H$. But b has vertex $\delta(D)$ and Lemma 13.7 gives us that no indecomposable summand of $kHtH$ has a vertex even containing $\delta(D)$.

We are now ready to prove Brauer's First Main Theorem. One way of motivating this is the question of the Green correspondent of the indecomposable $k[G \times G]$-module B, where B is a block of G. If D is a defect group of B then $N(D) \times N(D)$ contains the normalizer in $G \times G$ of $\delta(D)$, a vertex of B, so there is a corresponding $k[N(D) \times N(D)]$-module. It is a block, too!

Theorem 2 *If D is a p-subgroup of G and H is a subgroup of G containing $N(D)$ then there is a one-to-one correspondence between the blocks of H with defect group D and the blocks of G with defect group D given by letting the block b of H correspond with the block b^G of G.*

Of course, $C(D) \subseteq H$, so b^G makes sense, by the previous lemma. Once this is established, we shall speak of b as the *Brauer correspondent* of b^G in the most important case $H = N(D)$. Notice that b, as a $k[H \times H]$-module, is then the Green correspondent of the $k[G \times G]$-module b^G inasmuch as $b \mid (b^G)_{H \times H}$ while $\delta(D)$ is a vertex of b and b^G.

Proof Let b be a block of H with defect group D and set $B = b^G$. Now $H \times H$ contains the normalizer of $\delta(D)$ in $G \times G$, as $N(D) \times N(D) \subseteq H \times H$, b is an indecomposable $k[H \times H]$-module with vertex δD and $b \mid B_{H \times H}$ so Theorem 12.2 implies that B also has vertex δD, that is, B has defect group D. Moreover, that theorem implies that B and b are Green correspondents with respect to the subgroup $H \times H$. Hence, the map sending b to b^G is a one-to-one map of the blocks of H with defect group D to the blocks of G with defect group D. It remains to see that if B is a block of G with defect group D then it arises in this way. However, by Lemma 9.5, $B_{H \times H}$ has an indecomposable summand with vertex $\delta(D)$. But $B \mid kG$ and the only indecomposable summands of $(kG)_{H \times H}$ which have a vertex even containing $\delta(D)$ are the summands of kH, by Lemma 13.7, and these are just the blocks of H. Thus, there is a block b of H with defect group D such that $b \mid B_{H \times H}$ and so the proof is complete.

We are now going to establish the module version of Brauer's Second Main Theorem. The subject is the relation between the Green correspondence and the Brauer correspondence but we shall prove a more general result.

Theorem 3 *Let U be an indecomposable kG-module lying in the block B of G. Let V be an indecomposable kH-module, for a subgroup H of G, lying in the block b of H with vertex Q satisfying $C_G(Q) \subseteq H$. If $V \mid U_H$ then b^G is defined and $b^G = B$.*

Before proving this result let us derive a few consequences which illustrate its importance.

Corollary 4 *If* U *is an indecomposable* kG-*module lying in the block* B *and having vertex* Q *while* V *is the corresponding indecomposable* $kN_G(Q)$-*module which lies in the block* b *then* b^G *is defined and* $b^G = B$.

Roughly stated, modules which are in Green correspondence have their blocks correspond. This result is immediate from the theorem since $C_G(Q)$ is certainly contained in $N_G(Q)$ and $V \mid U_{N(Q)}$.

Corollary 5 *If* B *is a block of* G *with defect group* D *then there is an indecomposable* kG-*module lying in* B *with vertex* D.

To see this, let b be the block of $N(D)$ corresponding with B. Let S be a simple $kN(D)$-module lying in b so S is a $k[N(D)/D]$-module. Let V be an indecomposable projective module for $N(D)/D$ corresponding with S so, as S is a composition factor of V and V is indecomposable, we have that V also lies in b. Now V is a summand of the free module $k[N(D)/D]$ which is isomorphic with $(k_D)^{N(D)}$ so V is relatively D-projective and V_D is a direct sum of trivial kD-modules, since D being normal implies the same holds for $(k_D)^{N(D)}$. But then, by Lemma 9.5, V_D has a summand with the same vertex as V so V must have vertex D. Let U be the indecomposable kG-module corresponding with V under the Green correspondence. Hence, U also has vertex D and U lies in B, by Corollary 4, so Corollary 5 is proven.

Corollary 6 *The block* B *of* G *is a simple algebra if, and only if,* B *is of defect zero.*

Indeed, if B has defect group 1 then every indecomposable B-module has vertex 1, by Theorem 13.5, that is, every B-module is projective. Hence, all submodules of B-modules are summands so B is semisimple. But B is indecomposable as an algebra so B is simple as claimed. On the other hand, if B is simple then all B-modules are projective so Corollary 5 implies that B has defect group 1.

Let us now prove the theorem. There is a defect group of b containing Q, by Theorem 13.5, so its centralizer also is contained in H and Lemma 1 implies that b^G is defined. Hence, we need only assume that $b^G \neq B$ and argue to a contradiction. Let e be the unit element of b.

For any kH-module X we have $X = eX + (1-e)X$, a direct sum of kH-modules: indeed, eX is the canonical summand of X lying in b and, since $1-e$ is the sum of the unit elements of all the blocks of H other than b, $(1-e)X$ is a direct sum of the canonical summands lying in blocks of H other than b. If Y is a summand of X then $eY \mid eX$: if $X = Y + Z$ is a direct sum then $eX = eY + eZ$ is direct. Now $kG = kH + M$, where M is the sum of all the spaces $kHtH$, $t \notin H$, so $ekG = ekH + eM = b + eM$ is a direct sum of kH-modules. But $H \times 1 \subseteq H \times H$, b and M are $k[H \times H]$-modules and e commutes with all elements of kH so $ekG = b + eM$ is a direct sum of $k[H \times H]$-modules. Similarly, for $k[H \times H]$-modules, $eB \mid ekG$, so $eB \mid b + eM$ and $eM \mid M$. But b is an indecomposable $k[H \times H]$-module and is not a summand of $B_{H \times H}$, by assumption, so $eB \mid eM$ and $eB \mid M$. Thus, Lemma 13.7 implies that no indecomposable summand of the $k[H \times H]$-module eB has a vertex containing $\delta(Q)$.

It follows that no indecomposable summand of the $k[\delta(H)]$-module $eB_{\delta(H)}$ has a vertex containing $\delta(Q)$. Indeed, Mackey's Theorem implies that $eB_{\delta(H)}$ is relatively projective for the set of subgroups consisting of the conjugates of the vertices of the indecomposable summands of eB intersected with $\delta(H)$. We can now use the isomorphism $H \cong \delta(H)$ and transport of structure to deduce that eB, as a kH-module with elements of H acting by conjugation (so $h \cdot e\beta = he\beta h^{-1}$, $\beta \in B$, $h \in H$), has no indecomposable summand with a vertex containing Q. The same now holds for the kH-module $eB \otimes eU$, by exercise 9.2. Now $eV = V$ as V lies in b so $V \mid U_H$ implies that $V \mid eU$. Hence, in order to reach a contradiction, since V does have vertex Q, it suffices to show that $eU \mid eB \otimes eU$.

We first define a homomorphism φ of eU to $eB \otimes eU$ by setting

$$\varphi(w) = eE \otimes w$$

where $w \in eU$ and E is the unit element of B. This is clearly a linear transformation and it is a kH-homomorphism since whenever $h \in H$

$$\varphi(hw) = eE \otimes hw$$
$$= heEh^{-1} \otimes hw$$
$$= h(eW \otimes w)$$
$$= h(\varphi(w))$$

as e and E commute with h.

Next we define a linear transformation ψ from $eB \otimes eU$ to eU by

$$\psi(a \otimes w) = aw$$

for $a \in eB$ and $w \in eU$. Note that $aw \in eU$ as $a \in eB$ and that the bilinearity of the formula means the map is well defined. The map ψ is also a kH-

homomorphism. If $h \in H$ then

$$\psi(h \cdot a \otimes w) = \psi(hah^{-1} \otimes hw)$$
$$= hah^{-1}hw$$
$$= h(aw)$$
$$= h\psi(a \otimes w).$$

Finally, if $w \in eU$ then $ew = w$ and also $Ew = w$ since $w \in U$ which lies in B. Thus

$$\psi(\varphi(w)) = \psi(eE \otimes w)$$
$$= eEw$$
$$= w$$

so $eU \,|\, eB \otimes eU$.

Exercise

1 Let $G = SL(2, p)$, $H = N_G(P)$ for a Sylow p-subgroup P of G and determine the blocks of H and the blocks of G to which they correspond.

15 Canonical module

The First Main Theorem shows that the determination of the blocks of G with defect group D is a local question: it is enough to answer the question for $N(D)$. In this section we shall see that it is easy to describe the blocks of $N(D)$ with defect group D. To do this we need to develop some more Clifford theory as it pertains to blocks to be able to use it in studying $N(D)$.

Hence, we fix a normal subgroup N of G and a block B of G. If b is a block of N then its stabilizer, written Stab(b), is the subgroup of G consisting of all elements g of G such that $gbg^{-1} = b$. Since kN has a unique decomposition into blocks and since the elements of G induce automorphisms of the algebra kN by conjugation, the elements of G permute the blocks of N and the subgroup fixing b is Stab(b).

The way to connect the blocks of G and those of N is by means of the notion of covering. We say B *covers* b if there is a kG-module lying in B whose restriction to N has a summand lying in b. We shall see that this relation is quite restrictive and not at all as broad as it first might appear. Let us assume also that B does cover the block b. The next result contains the basic facts we required.

Theorem 1 *With the above assumptions, the following hold:*

(1) *The blocks of N covered by B form a single conjugacy class of blocks of N under conjugation by G;*

(2) *Each defect group of b is the intersection of a defect group of B with N;*

(3) *There is a defect group of B contained in* Stab(b);

(4) *There is a block B' of G covering b such that B' has a defect group D' containing a conjugate of a defect group of every block of G covering b and such that*

$$|D':D' \cap N| = |\text{Stab}(b):N|_p;$$

(5) *If the centralizer in G of a defect group of b is contained in N then B = b^G and B is the only block of G covering b.*

There are several comments called for. Notice that (2) does *not* claim that the intersection of each defect group of B with N is a defect group of b; these intersections, while all conjugate in G, may not be conjugate in N. The fourth assertion gives us a formula for the order of D', namely, the product of the order of a Sylow p-subgroup of Stab(b)/N and the order of a defect group of b, since $D' \cap N$ is conjugate in G with a defect group of b, by (2). This applies to b^G when (5) holds. Note that b^G is defined in (5) under the hypothesis of that assertion.

In order to establish the theorem, we shall first state and prove two lemmas. Let $b = b_1, \ldots, b_n$ be representatives of the conjugacy classes, under the action of G, of the blocks of N. Let $X = \delta(\text{Stab}(b)) \cdot N \times N$ so X is a subgroup of $G \times G$. Moreover, b is a subspace of the $k[G \times G]$-module kG and is invariant under the action of $X \subseteq G \times G$. Hence, we can regard b not just as a $k[N \times N]$-module but as a kX-module; we denote this module, with an obvious abuse of notation, as b_X.

Lemma 2 *With the above assumptions, the following hold:*

(1) *kG is the direct sum of the $k[G \times G]$-modules kGb_iG, $1 \leqslant i \leqslant n$;*

(2) *As a kN-module, kGbG is the canonical summand of kG lying in blocks of N conjugate with b;*

(3) *As a $k[N \times N]$-module, we have a direct sum decomposition*

$$kGbG = \sum_{t \in G \times G/X} tb;$$

(4) *As a $k[G \times G]$-module, $kGbG \cong (b_X)^{G \times G}$.*

Proof The module kGb_iG contains every conjugate of b_i: if $g \in G$ then $gb_ig^{-1} \subseteq kGb_iG$. Hence, the sum of all the kGb_iG, $1 \leqslant i \leqslant n$, contains all the blocks of H, that is, contains kH. Since the sum is closed under multiplication with G it must be all of kG. Once (2) is established we will have that the sum is direct, by Proposition 13.2.

In order to prove (2), we must show that

$$kGbG = \left(\sum_g gbg^{-1} \right) kG.$$

But $g_1 bg_2 = (g_1 bg_1^{-1})g_1 g_2$ so the left-hand side is contained in the right-hand one. On the other hand, $(gbg^{-1})kG = kgbG$, as $g^{-1}G = G$.

Turning to (3), each subspace tb is a $k[N \times N]$-submodule: if $y_1, y_2 \in N$ then

$$(y_1, y_2)tb = t \cdot t^{-1}(y_1, y_2)t \cdot b = tb$$

as b is a $k[N \times N]$-module. Moreover, $kGbG$ is the sum of these modules. Indeed, if $g_1, g_2 \in G$ then $(g_1, g_2) = tx$, for a representative t and $x \in X$, so

$$g_1 bg_2^{-1} = (g_1, g_2)b = txb = tb,$$

while $kGbG$ is the sum of all the spaces $g_1 bg_2^{-1}$. Finally, the sum is direct. Otherwise, a non-trivial linear combination, after multiplication with the inverse of a suitable one of the coset representative for X in $G \times G$, implies that

$$b \cap \sum_{(g_1, g_2) \notin X} (g_1, g_2)b \neq 0.$$

Hence, if $(g_1, g_2) \notin X$, it suffices to prove that either $(g_1, g_2)b$ is contained in $k(G - N)$ or it is contained in gbg^{-1} for some $g \in G$, $g \notin \text{Stab}(b)$; this is sufficient because kG is the direct sum of its subspaces kN and $k(G - N)$ while kN is the direct sum of its blocks. However, $(g_1, g_2)b = g_1 g_2^{-1}(g_2 bg_2^{-1})$ so if $g_1 g_2^{-1} \notin N$ then we are done. If $g_1 g_2^{-1} = y \in N$ then

$$\begin{aligned}
(g_1, g_2)b &= yg_2 bg_2^{-1} \\
&= g_2(g_2^{-1} yg_2)bg_2^{-1} \\
&= g_2 bg_2^{-1}.
\end{aligned}$$

If $g_2 \in \text{Stab}(b)$ then $(g_2, g_2) \in \delta(\text{Stab}(b))$ and $(g_1, g_2) = (g_1 g_2^{-1}, 1)(g_2, g_2) = (y, 1)(g_2, g_2) \in X$, a contradiction. Thus (3) is established.

The last statement is now immediate from our characterization of induced modules, inasmuch as (3) shows b_X generates the $k[G \times G]$-module $kGbG$ while the directness of the sum in (3) gives the required formula for the relation in the dimensions of $kGbG$ and b.

Lemma 3 *Under the above assumptions, the following are equivalent:*

(1) $b \mid B_{N \times N}$;

(2) $B \mid kGbG$;

(3) *If U is a kG-module lying in B then U_N is a direct sum of modules, one lying in each of the conjugates of b.*

Thus, the blocks covered by B are exactly the conjugates of b , no more, no less. To determine the blocks covered by B it is enough to examine the restriction of any one module lying in B to N .

Proof We shall show that the first two statements are equivalent and then that the last two are equivalent. Lemma 2 tells us that kG is the direct sum of its ideals kGb_iG , $1 \leqslant i \leqslant n$, and the uniqueness of the decomposition into blocks implies that $B \mid kGb_iG$, for some i . If (2) holds then, in particular, $bB \neq 0$ so $b(kGb_iG) \neq 0$. But, by the second part of Lemma 2, the only blocks of N which have a non-zero product with kGb_iG are the conjugates of b_i . Hence, b is conjugate with b_i , that is, $i = 1$, $b_i = b$ and (2) holds.

On the other hand, suppose that (2) is valid. We shall use the decomposition in part (3) of Lemma 2. Since $N \times N$ is certainly normal in $G \times G$, each of the modules tb is a conjugate of the $k[N \times N]$ -module and is therefore indecomposable. Since $B \mid kGbG$, the Krull–Schmidt theorem implies that $tb \mid B_{N \times N}$ for some t . Hence, b is a summand of the restriction to $N \times N$ of the $k[G \times G]$ -module obtained from B by conjugation by t^{-1} ; but that is isomorphic with B , as $t \in G \times G$, so (2) holds.

Again, if (2) holds, then $B \nmid kGb_iG$, $i > 1$, so $(kGb_iG)U = 0$ as U lies in B . In particular, $(gb_ig^{-1})U = 0$ for all $i > 1$, all g in G . On the other hand, $BU = U$, so $(kGbG)U = U$. Hence, there are g_1 and g_2 in G with $(g_1bg_2)U \neq 0$, that is, $g_1bU \neq 0$, as $g_2U = U$. Therefore, $bU \neq 0$ follows by premultiplication with g_1^{-1} . If g is in G then we also have $(gbg^{-1})U = gbU \neq 0$ as $bU \neq 0$. Thus, (3) is valid.

Finally, suppose that (3) holds. In particular, $bU \neq 0$ so $(kGbG)U \neq 0$. But $B \mid kGb_iG$ for some i , so $(kGb_jG)U = 0$ for all $j, j \neq i$; hence $i = 0$ and (2) is valid.

We now are able to establish Theorem 1.

Proof Lemma 3 has already established the first assertion of the theorem. It also implies the fifth statement. In fact, the hypothesis of (5) implies that b^G is defined so b^G is the unique block of G such that $b \mid (b^G)_{N \times N}$. Since B has this property, by Lemma 3 as it covers b , we have $B = b^G$ and the required uniqueness. The third statement is yielded by Lemma 2: it implies that B , as a $k[G \times G]$ -module, is relatively X -projective and so a vertex of B is contained in $\text{Stab}(b) \times \text{Stab}(s)$ and its projection on the first component, $\text{Stab}(b)$, is a defect group.

We have that $b \mid G_{N \times N}$ so, by Mackey's Theorem, the vertex of b is conjugate in $G \times G$ with a subgroup of the vertex of B. Hence, the defect group of b is conjugate in G with a subgroup of the defect group of B, as we may see by projecting on G, the first component of $G \times G$. On the other hand, by Exercise 9.6 and Lemma 9.7, there is an indecomposable summand of $B_{N \times N}$ which has a vertex containing the intersection of $N \times N$ and a vertex of B. However, by Lemma 2 and its proof, $B_{N \times N}$ is the direct sum of the indecomposable modules tb, each of which is isomorphic with a conjugate of b as $k[N \times N]$-modules. Hence, one and thus all of these modules has a vertex which contains the intersection of $N \times N$ and a vertex of B, so we have proved (2).

By Lemma 3, the blocks covering b are the indecomposable summands of $kGbG$ which is isomorphic with $(b_X)^{G \times G}$ by Lemma 2. Hence, each block, as a $k[G \times G]$-module, has a vertex contained in a vertex of b_X. But b_X is a summand of the restriction to X of $(b_X)^{G \times G}$ so there is a block B' of G covering b with $b_X \mid B'_X$. Hence, by Mackey's Theorem, a vertex of b_X is conjugate with a subgroup of the vertex of B'; thus, there is a vertex of B' which is a vertex of b_X and so B' satisfies the first part of (4). Moreover, if D' is a defect group of B', so $\delta(D')$ is a vertex of the $k[G \times G]$-module B', we may assume that $\delta(D') \subseteq X$. Since

$$|\delta(D'):\delta(D') \cap (N \times N)| = |D':D' \cap N|$$

and

$$|X:N \times N| = |\text{Stab}(b):N|,$$

in order to prove that (4) holds we need only show that if Q is a vertex of b_X then $|Q:Q \cap (N \times N)|_p = |X:N \times N|_p$. However, the restriction of b_X to $N \times N$ is just b, which is indecomposable, so we are done by Lemma 9.8.

We are now ready to apply our results to the situation of the First Main Theorem.

Theorem 4 *There is a one-to-one correspondence between blocks of G with defect group D and conjugacy classes in $N(D)$ of blocks β of $DC(D)/D$ of defect zero such that $|\text{Stab}(\beta):DC(D)|$ is not divisible by p.*

Hence $\text{Stab}(\beta)$ consists of all $x \in N(D)$ such that conjugation by x leaves β fixed; this makes sense since conjugation by elements of $N(D)$ induces automorphisms of the algebra $k[DC(D)/D]$. The blocks of $DC(D)/D$ of defect zero are exactly the matrix algebra summands of the group algebra

$k[DC(D)/D]$ and so these correspond one-to-one with the projective simple modules for $k[DC(D)/D]$. Thus, in order to determine the blocks of G with defect group D we need to find the projective simple $k[DC(D)/D]$-modules on which $N(D)$ acts suitably by conjugation. The simple projective module that the proof of this theorem gives us in correspondence with B is called the *canonical module* of B (even though it is not a B-module but is a $k[DC(D)/D]$-module). We shall prove this theorem by establishing two lemmas.

Lemma 5 *There is a one-to-one correspondence between the blocks of G with defect group D and conjugacy classes in $N(D)$ of blocks β of $DC(D)$ with defect group D such that $|\text{Stab}(\beta):DC(D)|$ is not divisible by p.*

Proof If B is a block of G with defect group D and b is the corresponding block of $N(D)$ then let β be a block of $DC(D)$ covered by b. Since D is certainly normal in $DC(D)$ it is contained in a defect group of β so the centralizer of that defect group is contained in $DC(D)$. Therefore, $\beta^{N(D)}$ is defined, it equals b and b is the unique block of $N(D)$ covering β. Moreover, β has defect group D: since $\beta^{N(D)} = b$ a defect group of β, which must contain D, is contained in a defect group of b, by Lemma 14.1. Since b has defect group D we also have, by part (4) of Theorem 1, that p does not divide $|\text{Stab}(\beta):DC(D)|$. Hence, we have a one-to-one map from blocks of G with defect group D to conjugacy classes of blocks of $DC(D)$ with the required properties.

Suppose next that β is a block of $DC(D)$ with defect group D such that $|\text{Stab}(\beta):DC(D)|$ is not divisible by p. Thus β^G is the unique block of $N(D)$ covering β and, by parts (2) and (4) of Theorem 1, a defect group of β^G, which must contain the defect group D of β, has order $|\text{Stab}(\beta):DC(D)|_p|D|$. Thus, b has defect group D and so, by the First Main Theorem, is the block of $N(D)$ corresponding with a block of G with defect group D. Hence, the lemma is proved.

The next lemma is more general than we require here. It now immediately implies Theorem 4.

Lemma 6 *There is a one-to-one correspondence between blocks of $DC(D)$ and blocks of $DC(D)/D$ such that if a block of $DC(D)$ has defect group P then the corresponding block of $DC(D)/D$ has defect group P/D. Moreover, this correspondence is compatible with conjugation by elements of $N(D)$.*

Proof Since D is a normal p-subgroup of $DC(D)$, every simple $k[DC(D)]$-

module has D acting trivially and so is also a $k[DC(D)/D]$-module. We claim that two simple modules are in the same block of $DC(D)$ if, and only if, they are in the same block of $DC(D)/D$ when considered as $k[DC(D)/D]$-modules.

Indeed, by Proposition 13.3, it suffices to establish the following assertion: if S and T are non-isomorphic simple $k[DC(D)]$-modules and there exists a non-split exact sequence

$$0 \to T \to U \to S \to 0$$

of $k[DC(D)]$-modules then D acts trivially on U. However, $S_{C(D)}$ and $T_{C(D)}$ are non-isomorphic simple modules as D acts trivially on S and on T. Thus, $U_{C(D)}$ has exactly two composition factors and these are non-isomorphic. Hence $U_{C(D)}$ is either uniserial or semisimple. Moreover, in the first case, the one non-zero proper submodule of $U_{C(D)}$ is not isomorphic with the one non-zero proper homomorphic image of $U_{C(D)}$ so every endomorphism of $U_{C(D)}$ is an isomorphism, that is, $\text{End}(U_{C(D)})$ is local and has no non-zero nilpotent elements so $\text{End}(U_{C(D)}) \cong k$. In the second case $\text{End}(U_{C(D)}) \cong k + k$ so $\text{End}(U_{C(D)})$ is always semisimple. However, elements of D commute with elements of $C(D)$ so the linear transformations of U induced by the action of elements of D are in $\text{End}(U_{C(D)})$. Hence, we have a homomorphism of the algebra kD into the algebra $\text{End}(U_{C(D)})$. But kD is a local algebra and each subalgebra of $\text{End}(U_{C(D)})$ is semisimple, by the above analysis, so $\text{rad}(kD)$ is the kernel of this algebra homomorphism. In particular, if $d \in D$ then $d - 1 \in \Delta(kD) = \text{rad}(kD)$ so $d - 1$ induces the zero linear transformation on U and the claim is valid.

Hence, if b is a block of $DC(D)$ there is a corresponding block $\bar{b}$ of $DC(D)/D$, the one having the same simple modules lying in it. Since a module lies in a block exactly when its composition factors do, we have that the $\bar{b}$-modules are exactly the b-modules on which D acts trivially.

Suppose that P is the defect group of b. We shall first show that each indecomposable $\bar{b}$-module has a vertex contained in P/D; this immediately implies that $\bar{b}$ has defect group contained in P/D, by Theorem 13.5 and Corollary 14.5. Indeed, if U is a $\bar{b}$-module then U is a b-module and so is relatively P-projective and so $U | (U_P)^{DC(D)}$. But D acts trivially on U so we have $U | (U_{P/D})^{DC(D)/D}$ so U is relatively P/D-projective as a $\bar{b}$-module.

Finally, it is clear now that the lemma will be entirely proven, once we show that P/D is the vertex of an indecomposable $\bar{b}$-module. To do this we shall apply the Brauer correspondence and the Green correspondence *within* $DC(D)$. Keeping this in mind, let β be the block of $N(P)$ (that is, $N_{DC(D)}(P)$) which is the Brauer correspondent of b. We argue just as in the proof of Corollary 14.5: Let V be an indecomposable projective $kN(P)/P$-module lying in β so V has vertex P and has trivial restriction to P; the Green correspondent U of V lies in b, has vertex P and $U \mid V^{DC(D)}$. The latter assertion implies that D acts trivially on U so U, in fact, lies in $\bar{b}$. Finally, U, as a $\bar{b}$-module, also has vertex P: if $U \mid (U_{Q(D)})^{DC(D)/D}$ for a subgroup Q of D then $U \mid (U_Q)^{DC(D)}$.

Exercises

1   If K is a non-identity normal subgroup of order not divisible by p in G then G has more than one block.

2   If Q is a normal p-subgroup of G and $Q \supseteq C(Q)$ then G has exactly one block.

3   With the notation of Theorem 1, if $g_1, g_2 \in G$ and $(g_1, g_2)b = b$ (regarding b as contained in the $k[G \times G]$-module kG) then $(g_1, g_2) \in X$.

16 Subpairs

We have expended considerable effort describing the structure of kG-modules in terms of the local subgroups. The Green correspondence gives a determination of the non-projective indecomposable kG-modules and the Brauer correspondence does the same for blocks of positive defect. In this section we shall see how the map sending a block b of a p-local subgroup to the block b^G of G can be described locally. More specifically, we shall see how to find the Brauer correspondent of b^G and pass from b to the defect group D of b^G and the corresponding block of b^G.

There is one very important special case in which the result is also very easy to understand. The *principal block* of $b_0(G)$ is the block in which the trivial kG-module k lies. Generally, this block is the most useful one and has the most complicated structure. Since the vertex of the trivial module is a Sylow p-subgroup, it follows that $b_0(G)$ has the Sylow p-subgroups as its defect groups. If N is a normal subgroup of G then $b_0(G)$ only covers $b_0(N)$ since the restriction of k to N is the trivial kN-module. However, the most important property of $b_0(G)$ is given by Brauer's Third Main Theorem. This answers the opening question of the section for $b_0(G)$.

Theorem 1 *If b is a block of the subgroup H of G, D is a defect group of b and $C_G(D) \subseteq H$ then $b^G = b_0(G)$, if, and only if, $b = b_0(H)$.*

Proof First, suppose that $b = b_0(H)$. We are going to apply Theorem 14.3 to determine b^G. The trivial kG-module k lies in $b_0(G)$, its restriction k_H to H lies in b and its vertex is a Sylow p-subgroup of H. But b has the Sylow p-subgroups as its defect groups so, by assumption, the centralizer in G of a vertex of k_H lies in H. Hence, by 14.3, $b^G = b_0(G)$.

Second, assume that $b \neq b_0(H)$ but that $b^G = b_0(G)$; we must argue to a contradiction. Let b' be the block of $N_H(D)$ corresponding with b under the Brauer correspondence, applied to the group H, so b' has defect group D as well and $(b')^H = b$. But $C(D) \subseteq H$ so $(b')^G$ is defined and therefore $(b')^G = b^G = b_0(G)$ by Lemma 14.1. If $b' = b_0(N_H(D))$ then $b = b_0(H)$, by the preceding paragraph, so we have $b' \neq b_0(N_H(D))$. Let β be a block of $DC_H(D) = DC(D)$ covered by b' so, as usual, β has defect group D and $(\beta)^{N_H(D)} = b'$, so, again by the above, $\beta \neq b_0(DC(D))$. Moreover, β^G is defined so $\beta^G = (b')^G = b_0(G)$. At this point we shall no longer have any use for the subgroup H but shall use β to reach the desired contradiction.

Therefore, we have that D is a p-subgroup of G, β is a block of $DC(D)$ with defect group D, $\beta \neq b_0(DC(D))$ but $\beta^G = b_0(G)$. We shall proceed, by 'downward' induction on the order of D, to reach the desired contradiction. Suppose first that D is a Sylow p-subgroup of G. In particular, $b_0(G) = \beta^G = (\beta^{N(D)})^G$. But $b_0(N(D))^G = b_0(G)$, by the first paragraph of the proof, so the First Main Theorem implies that $\beta^{N(D)} = b_0(N(D))$. But, by Theorem 15.1, $\beta^{N(D)}$ is the unique block of $N(D)$ covering β, while $b_0(N(D))$ only covers $b_0(DC(D))$ so $\beta_0 = b_0(DC(D))$, a contradiction. Hence, we may now assume that if E is a p-subgroup of G properly containing D, γ is a block of $EC(E)$ with defect group E and $\gamma^G = b_0(G)$ then $\gamma = b_0(EC(E))$.

By Lemma 14.1, $\beta^{N(D)}$ has a defect group E containing D. We claim that it contains D properly: if $D = E$ then $(\beta^{N(D)})^G = \beta^G = b_0(G)$ yields that D is a Sylow p-subgroup, by the First Main Theorem. By Lemma 15.5 applied to the group $N(D)$, we have that there is a block γ of $EC_{N(D)}(E) = EC(E)$ (as $C(E) \subseteq C(D)$) with defect group E and $\gamma^{N(D)} = \beta^{N(D)}$. Therefore,

$$\gamma^G = (\gamma^{N(D)})^G = (\beta^{N(D)})^G = \beta^G = b_0(G)$$

so, by our induction, $\gamma = b_0(EC(E))$. Hence, $\gamma^{N(D)} = b_0(N(D))$. But $\beta^{N(D)} =$ The proof relies on the use of *Brauer pairs*, that is, pairs (Q, b), where Q is a p- we have a contradiction.

The proof relies on the use of *Brauer pairs*, that is, pairs (Q, b), where Q is a p-subgroup of G and b is a block of $QC(Q)$ with defect group Q. With arduous effort, Brauer used these pairs as a computational device, with great effect. However, much deeper and simpler things are going on: there is a synthesis between the basic results of block theory and the most basic results of group theory. The idea is to introduce a more general type of pair and show they can be treated as a sort of 'generalized' subgroup. In particular, we shall be able to tie together Sylow's theorem and the First Main Theorem. This leads to two things: tremendous simplifications; a way to generalize results from local group theory to block theory.

If Q is a p-subgroup of G and b_Q is a block of $QC(Q)$ then the pair (Q, b_Q) is called a *subpair* (for 'subgroup block pair') of G. We make no restriction on the defect groups of b_Q. To each subpair corresponds a block of G: Q is normal in $QC(Q)$ and is therefore contained in a defect group of b_Q so $b_Q^G = b$ is defined. In this case, we say that (Q, b_Q) is a *b-subgroup* of G; it is a subgroup together with the additional structure of a selected block. The idea is to treat the totality of b-subgroups of G in the same way we handle all p-subgroups, to make believe that 'b' is a prime. This can be carried out and we shall establish some of the basic results, and, in so doing, answer the question that opened the section.

We begin by defining normality for subpairs. If (Q, b_Q) and (R, b_R) are subpairs then we say that (R, b_R) is *normal* in (Q, b_Q) and write $(R, b_R) \lhd (Q, b_Q)$ provided the following conditions are satisfied:

(1) $R \lhd Q$;
(2) b_R is invariant under conjugation by Q;
(3) $(b_Q)^{QC(R)} = b_R^{QC(R)}$.

Thus, Q leaves R and b_R invariant under conjugation. The third condition, which makes sense as $QC(R)$ is a subgroup, implies that $b_Q^G = b_R^G$ so that if this is the block b of G then (Q, b_Q) and (R, b_R) are both b-subgroups of G. But the third condition says that b_Q and b_R are closely related in a specific and useful way.

We shall now define containment for subpairs and we do this in analogy with the situation for p-subgroups: the p-subgroup R is contained in the p-subgroup Q if, and only if, R is subnormal in Q. Hence, if (Q, b_Q) and (R, b_R) are subpairs then we say that (R, b_R) is *contained* in (Q, b_Q) and write $(R, b_R) \subseteq (Q, b_Q)$ provided there are subpairs (R_i, b_i), $1 \leqslant i \leqslant n$, with $(R, b_R) = (R_1, b)$, $(Q, b_Q) = (R_n, b_n)$ and $(R_i, b_i) \lhd (R_{i+1}, b_{i+1})$, $1 \leqslant i < n$. In particular, $b_i^G = b_{i+1}^G$, $1 \leqslant i < n$, so all the subpairs involved, including (Q, b_Q) and (R, b_R) are b-subgroups for a block b of G.

A very important definition is the next one: if b is a block of G then a b-subgroup (Q, b_Q) is a *Sylow b-subgroup* if Q is a defect group of b. This is justified by a fundamental result.

Theorem 2 *If b is a block of G then every b-subgroup is contained in a Sylow b-subgroup and all the Sylow b-subgroups are conjugate.*

This answers the question we began this section with. Indeed, let β be a block of $L = N(R)$, R a non-identity p-subgroup. If β has defect group Q then, as above, there is a block b_Q of $QC_L(Q) = QC(Q)$ (as $R \subseteq Q$) such that $b_Q^L = \beta$ so $\beta_Q^G = \beta^G$. The proof of this theorem will show how to find, in terms of local structure only, a Sylow subpair containing (Q, b_Q). If this is (D, b_D) then $b_D^{N(D)}$ is the Brauer correspondent of β^G.

Proof Let us first establish the conjugacy assertion. Since all the defect groups are conjugate, it suffices to prove that (D, b_1) and (D, b_2) are conjugate where D is a defect group of b and each (D, b_i) is a b-subgroup. But $b_i^{N(D)}$ is a block of $N(D)$ and $(b_i^{N(D)})^G = b_i^G = b$. Since b has defect group D and since each defect group of b_i must contain D, Lemma 14.1 yields that b_i and $b_i^{N(D)}$ each have defect group D. But $(b_1^{N(D)})^G = b = (b_2^{N(D)})^G$ so the First Main Theorem implies that $b_1^{N(D)} = b_2^{N(D)}$. Hence, the results of the last section imply that b_1 and b_2 are covered by the same block of $N(D)$ and are thus conjugate in $N(D)$, as required.

Therefore, it remains to show that if (Q, b_Q) is a maximal b-subgroup then Q is a defect group of b. Let $H/QC(Q)$ be a Sylow p-subgroup of $\mathrm{Stab}(b_Q)/QC(Q)$. Applying results of the previous section to the normal subgroup $QC(Q)$ of the group H, we deduce that b_Q^H is the unique block of H covering b_Q and that it has a defect group E with $EQC(Q) = H$ and $E \cap QC(Q) = R$, a defect group of b_Q. Also from results of the preceding section, there is a block b_E of $EC_H(E) = EC(E)$ (as $E \supseteq R \supseteq Q$) such that $b_E^H = b_Q^H$. But $H = EC(Q)$ so $b_E^{EC(Q)} = b_Q^{EC(Q)}$, Q is normal in E and E stabilizes b_Q; that is, $(Q, b_Q) \lhd (E, b_E)$. Our maximality assumption implies that $Q = E$ so $Q = R = E$ and $|\mathrm{Stab}(b_Q): QC(Q)|_p = 1$. Again using Theorem 15.1, we have that $b_Q^{N(Q)}$ is the only block of $N(Q)$ covering b_Q and that it has defect group Q. Hence by the First Main Theorem, $(b_Q^{N(Q)})^G = b_Q^G = b$ also has Q as a defect group.

The next result shows that there are many b-subgroups and gives a useful uniqueness statement.

Theorem 3 *If (P, b_P) is a subpair and Q is a subgroup of P then there is a unique block b_Q of $QC(Q)$ such that $(Q, b_Q) \subseteq (P, b_P)$. Moreover, if Q is normal in P then $(Q, b_Q) \lhd (P, b_P)$.*

Before proving this result, let us see what it tells us about the principal block. Let Q be a p-subgroup of G so $(Q, b_0(QC(Q)))$ is a subpair, in fact it is a $b_0(G)$-subgroup since $(b_0(QC(Q)))^G = b_0(G)$. Moreover, if b_Q is a block of $QC(Q)$ with (Q, b_Q) a $b_0(G)$-subgroup then $b_Q = b_0(QC(Q))$, by Theorem 1.

Hence, there is a one-to-one correspondence between p-subgroups of G and $b_0(G)$-subgroups of G. Moreover, if P is another p-subgroup of G then $Q \subseteq P$ if, and only if $(Q, b_0(QC(Q))) \subseteq (P, b_0(PC(P)))$. Indeed, by Theorem 3, there is a subpair (Q, b_Q) contained in $(P, b_0(PC(P)))$ so $b_Q^G = (b_0(PC(P)))^G = b_0(G)$ and it must be that $b_Q = b_0(QC(Q))$. It is also now clear from Theorem 3 that $(Q, b_0(QC(Q))) \lhd (P, b_0(PC(P)))$ if, and only if, $Q \lhd P$. The study of p-subgroups is exactly the same as the study of the $b_0(G)$-subgroups of G and we have generalized the idea of p-subgroups.

In order to prove the theorem we first establish a preliminary result.

Lemma 4 *Let (P, b_P) be a subpair of G.*

(1) If Q is a normal subgroup of P then there exists a unique block b_Q of $QC(Q)$ such that $(Q, b_Q) \lhd (P, b_P)$.

(2) If Q and R are subgroups of P with $R \subseteq Q$ and (Q, b_Q) and (R, b_R) are subpairs normal in (P, b_P) then $(R, b_R) \lhd (Q, b_Q)$.

Proof Under the first hypothesis, $QC(Q)$ is a normal subgroup of $PC(Q)$ so there is a block b_Q of $QC(Q)$ covered by the block $b_P^{PC(Q)}$ of $PC(Q)$. But, by Theorem 15.1, $b_Q^{PC(Q)}$ is the unique block of $PC(Q)$ covering b_Q so it must equal $b_P^{PC(Q)}$. Hence, in order to prove that $(Q, b_Q) \lhd (P, b_P)$ we need only show that P stabilizes b_Q. But P is contained in a defect group of b_P so it is contained in a defect group of $b_P^{PC(Q)}$. But this is $b_Q^{PC(Q)}$ so, again by Theorem 15.1, a conjugate of P in $PC(Q)$ is contained in the stabilizer Stab(b_Q) of b_Q in $PC(Q)$. But $C(Q)$ certainly stabilizes b_Q so $P \subseteq$ Stab(b_Q), as desired.

In particular, Stab$(b_Q) = PC(Q)$ so b_Q is the only block of $QC(Q)$ covered by $b_Q^{PC(Q)}$, as it has no other conjugates in $PC(Q)$. In order to prove the uniqueness stated in (1), let (Q, b_Q') be another subpair which is normal in (P, b_P). Thus, $(b_Q')^{PC(Q)} = b_P^{PC(Q)}$ so $b_P^{PC(Q)}$ is the block of $PC(Q)$ covering b_Q' so $b_Q = b_Q'$.

The hypotheses of (2) are such that in order to establish the statement it is only necessary to prove that $b_Q^{QC(R)} = b_R^{QC(R)}$. But these are blocks of $QC(R)$ and they are stable in $PC(R)$ since $QC(R)$ is normal in $PC(R)$ and since P stabilizes b_Q and b_R by hypothesis. Hence, by Theorem 15.1, applied to the normal subgroup $QC(R)$ of $PC(R)$, it suffices to prove that

$$(b_Q^{QC(R)})^{PC(R)} = (b_R^{QC(R)})^{PC(R)}.$$

However,

$$(b_R^{QC(R)})^{PC(R)} = b_R^{PC(R)} = b_P^{PC(R)}$$

since $(R, b_R) \lhd (P, b_P)$. Moreover,

$$(b_Q^{QC(R)})^{PC(R)} = b_Q^{PC(R)}$$
$$= (b_Q^{PC(Q)})^{PC(R)}$$
$$= (b_P^{PC(Q)})^{PC(R)}$$

as $(Q, B_Q) \lhd (P, b_P)$ so we finally get

$$(b_Q^{QC(R)})^{PC(R)} = b_P^{PC(R)}$$

and the lemma is proven.

We conclude by proving Theorem 3. The last assertion of the theorem has already been shown to hold in the preceding lemma. If Q is a subgroup of P then Q is subnormal in P so repeated use of part (1) of Lemma 4 establishes the existence of a block b_Q of $QC(Q)$ with $(Q, b_Q) \subseteq (P, b_P)$. All that remains to be demonstrated is the uniqueness.

We proceed by induction on the index $|P:Q|$, the case of index one being trivial. Suppose that (Q, b_Q') is also a subpair contained in (P, b_P). From the definition of containment of subpairs, there is a proper normal subgroup R of P containing Q and a subpair (R, b_R) such that $(Q, b_Q) \subseteq (R, b_R) \lhd (P, b_P)$. Similarly, there is a proper normal subgroup R' and a subpair $(R', b_{R'})$ with $(Q, b_Q') \subseteq (R', b_{R'}) \lhd (P, b_P)$. Set $S = R \cap R'$ so S is also normal in P and so, by the preceding lemma, there is a subpair (S, b_S) normal in (P, b_P). By the second part of that lemma, we have $(S, b_S) \lhd (R, b_R)$ and $(S, b_S) \lhd (R', b_{R'})$. Since we have already proved the existence part of the theorem, there is a subpair $(Q, b_Q'') \subseteq (S, b_S)$. Hence, (Q, b_Q) and (Q, b_Q'') are contained in (R, b_R), so they are equal, by induction, and similarly (Q, b_Q') and (Q, b_Q'') are equal as they are contained in $(R', b_{R'})$.

It is interesting to note that the last argument traces back to an old proof of the Jordan–Hölder theorem, which proves that theorem directly without the Schreier refinement theorem.

Exercises

1 Determine the canonical module for the principal block of G.

2 If Q is a subgroup of the defect group D of the block B of G then there is an indecomposable kG-module U lying in B with Q a vertex of U.

3 The subpair (Q, b_Q) is a Sylow b_Q^N-subgroup (where $N = N((Q, b_Q)) = N(Q) \cap \text{Stab}(b_Q)$) if, and only if, (Q, b_Q) is a Sylow b_Q^G-subgroup.

V
Cyclic blocks

We have now amassed the ideas needed to develop one of the deepest and most important parts of representation theory: the structure theory of blocks with cyclic defect groups. We have already looked at groups with cyclic normal Sylow p-subgroups and at SL$(2, p)$ and achieved complete information about the structure of projective modules; we shall now do the same for any block with a cyclic defect group. This theory is one of the triumphs of the subject and, at the same time, the biggest challenge, in that the main problem is to generalize these results to theorems about arbitrary blocks.

17 Brauer trees

The structure of a block with a cyclic defect group is remarkably determined by the properties of a graph which is associated with the block. In fact, it is not just a graph, but a Brauer tree, which is a graph with some additional structure. A *Brauer tree* is defined as a finite, connected, undirected graph without loops or cycles (so it is a tree) together with two additional structures: a circular ordering of the edges emanating from each vertex; the assignment of a positive integer, called the *multiplicity*, to one of the vertices, called the *exceptional vertex*.

This requires some elucidation. The property of being a tree means there is no path in the graph, starting and ending at the same vertex but passing over a set of distinct edges, each just once. The ordering means that to each edge E emanating from a vertex v there is a 'next' edge also emanating from v, and a 'next' and so on, until each edge emanating from v has been 'counted' exactly once, in which case the edge E is 'next'. Notice that E also emanates from another vertex w, its other 'endpoint', so it is also involved in another ordering, the one for edges emanating from w. The ordering may be quite trivial, but is still relevant: if E and F are the only edges emanating from v then F is next after E and E is next after F.

There is a convenient way to picture all this. We draw the Brauer tree embedded in the plane so that the circular orderings are each just the counter-clockwise ordering given by 'counting' the edges emanating from a vertex by passing from one to the next one in a counter-clockwise manner. In fact, it is true and easy to prove that, since an embedding is possible, so one can think of a Brauer tree as a tree, embedded in the plane, together with a multiplicity attached to an exceptional vertex. To make it clear which vertex is exceptional, it is customary to draw the exceptional vertex as a blackened circle and leave all the other vertices as open circles.

There are two very important Brauer trees, in that they arise often. First, there is the *star* with exceptional vertex in the center, that is,

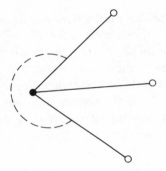

and there is the *open polygon*, here drawn with exceptional vertex at the end:

Now we wish to relate these trees to the structure of algebras. An algebra A is called a *Brauer tree algebra* if there is a Brauer tree so that the indecomposable projective A-modules can be described by a particular algorithm using the Brauer tree, one that we shall now enunciate. First, there is a one-to-one correspondence between the edges of the tree and the isomorphism classes of simple A-modules. We can picture this by labelling the edges of the tree, as drawn in the plane, by the corresponding simple A-modules. Second, if S is a simple A-module and P_S is the corresponding indecomposable projective A-module then $P_S \supseteq \mathrm{rad}(P_S) \supseteq \mathrm{soc}(P_S) \cong S$ and $\mathrm{rad}(P_S)/\mathrm{soc}(P_S)$ is the direct sum of two uniserial modules (each possibly the zero module) described by the orderings of the edges emanating from each of the endpoints of the edge corresponding to S, in a manner which we shall now describe.

Suppose that the edge labelled by S has the vertices v, w as its endpoints. Suppose that the edges emanating from v are labelled by $T_1, \ldots, T_r$ where T_1 is next after S, T_2 follows T_1 and so on, with all the T_i distinct, until S follows T_r. And suppose that the edges emanating from w are $U_1, U_2, \ldots, U_n$, similarly given so we have the following picture:

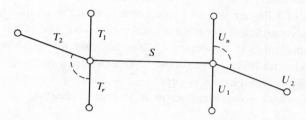

We are assuming that neither v nor w is the exceptional vertex; we shall have to make a modification if one of them is the exceptional vertex. Our final condition on A is that $\operatorname{rad}(P_S)/\operatorname{soc}(P_S)$ be the direct sum of two uniserial modules, one which has composition factors $T_1, \ldots, T_r$ and occurring in that order and the other with composition factors $U_1, \ldots, U_n$ occurring in that order. Since $\operatorname{rad}(P_S)$ is the unique maximal submodule of P_S and $\operatorname{soc}(P_S)$ is the unique minimal submodule of P_S (as it is simple, by assumption) we have a good picture of P_S:

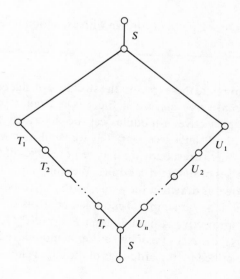

Here the nodes denote submodules (for example, P_S, rad(P_S) at the top) and the labels give the isomorphism class of successive quotients.

We need to describe the modification if a vertex is exceptional: assume v is exceptional with multiplicity m. All that we need to do is change the description of the first uniserial module, the one involving the T_i. It is now uniserial with the following composition factors and in the following order:

$$T_1, T_2, \ldots, T_r, S, T_1, \ldots \quad \ldots, S, T_1, \ldots, T_r.$$

Here, S occurs $m-1$ times, each T_i occurs m times. (Thus, if $m=1$ there is no difference with the non-exceptional case.)

One way to remember this recipe is to think along the following lines. To describe the uniserial corresponding to v, in describing P_S, imagine standing on the edge labelled S and taking a walk, in a counter-clockwise fashion, around the vertex v, going around m times, with $m=1$ if v is not the exceptional vertex, and keep track of the labellings of the edges crossed, *after* leaving the edge labelled S and *before* returning to that edge for the final time. The sequence of simple modules is the one we have described above.

Now let us examine some special cases, to help understand the concept and tie up with things we have done. Suppose A is a Brauer tree algebra for the Brauer tree

where the multiplicity is m. Thus, A has exactly one simple module S. This time the uniserial module for the right-hand vertex is the zero module; the one for the left-hand one is uniserial of composition length $m-1$. Thus, P_S is uniserial (remember what we said about rad(P_S) and soc(P_S)) of composition length $m+1$. We have seen such algebras. Let G be a cyclic p-group of order p^n: we have proved that kG is a Brauer tree algebra for the above tree with multiplicity p^n-1.

Next, suppose A is a Brauer tree algebra for the following Brauer tree, with multiplicity two, and edges labelled:

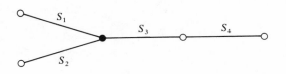

The projective indecomposables then look as follows:

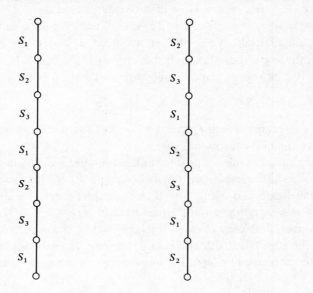

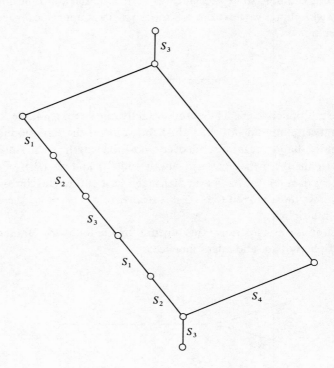

Let us describe two more cases where Brauer trees are relevant. First, suppose that P is a cyclic normal Sylow p-subgroup of G. We know from section 5 that there is a one-dimensional kG-module W very relevant to the structure of the indecomposable projective kG-modules. We also know, from exercise 13.3, that if B is a block of kG then we can list the simple modules lying in B, say, $S_1, \ldots, S_r$, in such a way that $S_i \otimes W \cong S_{i+1}, 1 \leq i < r$, and $S_r \otimes W \cong S_1$. Combining these facts, we have, in fact, that B is a Brauer tree algebra for a Brauer tree which is a star with r edges and exceptional vertex in the center with multiplicity $(p^n - 1)/r$, where P has order p^n, and the labelling is as follows:

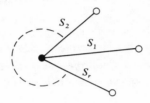

Second, suppose that $G = \mathrm{SL}(2, p)$. The simple modules $V_1, \ldots, V_{p-1}$ fall into two blocks and these blocks are Brauer tree algebras with the following descriptions (where $\varepsilon = \pm 1$ and $p \equiv \varepsilon$ (modulo 4)):

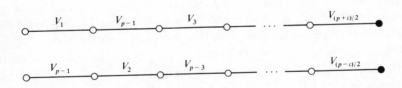

The multiplicity is two.

In the rest of the section we shall describe the main results of this chapter and just prove a single theorem.

We fix some notation. Let B be a block of G with defect group D cyclic of order $p^n, n \geq 1$. Let b be the block of $N(D)$ that is the Brauer correspondent of B. The index of $DC(D)$ in the stabilizer (in $N(D)$) of a block of $DC(D)$ covered by b is called the inertial index of B and we denote it by e. Since all the blocks of $DC(D)$ covered by b are conjugate in $N(D)$, it is well defined. The main result of this chapter is as follows:

Theorem 1 *The block B is a Brauer tree algebra for a tree with e edges and multiplicity $(p^n - 1)/e$.*

The proof proceeds in a number of stages, which we shall now describe. Let D_1 be the subgroup of order p in D and let $N_1 = N(D_1)$. Let $b_1 = b_1^{N_1}$ so b_1 is a block of N_1 having defect group D and $b_1^G = B$. In the next two sections we shall study the blocks of $C(D_1)$ covered by b_1 and then b_1 itself and prove the next result.

Theorem 2 *The block b_1 is a Brauer tree algebra for a star with e edges and exceptional vertex in the center with multiplicity $(p^n - 1)/e$.*

The basic result that allows us to pass from the structure of b_1 to that of B is very much like the results of section 10 on trivial intersections.

Theorem 3 *There is a one-to-one correspondence between isomorphism classes of indecomposable and non-projective B-modules and isomorphism classes of indecomposable and non-projective b_1-modules such that if U and V are B and b_1-modules, respectively, which correspond, then*

$$V^G \cong U \oplus Q,$$

$$U_{N_1} \cong V \oplus W$$

where Q is a projective kG-module and W is a direct sum of a projective kN_1-module and modules lying in blocks of N_1 other than b_1.

Proof We use the Green correspondence for the subgroup N_1 and p-subgroup D, which we may do as $N_1 \supseteq N$. We must first calculate the collections $\mathfrak{x}$ and $\mathfrak{y}$ of subgroups for this particular situation. First, suppose that $E = D \cap gDg^{-1} \neq 1$ for some $g \in G$. Thus, E is a non-identity subgroup of D and of gDg^{-1} so E contains D_1 and gD_1g^{-1}, the unique minimal subgroups of D and gDg^{-1}, respectively. But E is also cyclic so contains a unique subgroup of order p so $D_1 = gD_1g^{-1}$, that is, $g \in N_1$. Hence, $\mathfrak{x} = \{1\}$. Moreover,

$$\mathfrak{y} = \{N_1 \cap gDg^{-1} \mid g \in G - N\},$$

so we similarly get that the non-identity subgroups of groups that lie in $\mathfrak{y}$ are not conjugate in N_1 with a subgroup of D. Hence, any relatively $\mathfrak{y}$-projective kN_1-module is the direct sum of indecomposable modules which are projective (have vertex 1) or lie in blocks of N_1 other than b_1 since their non-identity vertices are not conjugate in N_1 with a subgroup of a defect group of b_1.

Let U be an indecomposable and non-projective kG-module lying in B so its vertex is a non-identity subgroup of D. Therefore, it has a Green correspondent V which is an indecomposable kN_1-module with the same vertex. If V lies in the block b' of N_1 then $(b')^G = B$ since $C(D_1) \subseteq N_1$ guarantees also that the centralizer of the vertex of V also is contained in N_1. Therefore, if D' is the defect group of b' then $D' \supseteq D_1$, as D' contains the vertex of V, and D' is cyclic as it is conjugate in G with a subgroup of D. Hence, $N(D') \subseteq N(D_1) = N_1$ so the First Main Theorem applies so $D' = D$ and $b' = b_1$. Hence, by the Green correspondence, $U_{N_1} \cong V \oplus W$ as required, since V lies in b_1 and W is relatively $\mathfrak{y}$-projective.

On the other hand, if V is a non-projective indecomposable kN_1-module lying in b_1 then N_1 contains the centralizer of its defect group, a non-identity subgroup of D, so the Green correspondent U of V lies in $b_1^G = B$. Moreover, $V^G \cong U + Q$, as claimed, since $\mathfrak{x} = \{1\}$. The one-to-one correspondence is now immediate since $V \mid (V^G)_{N_1}$ and $U \mid (U_{N_1})^G$, the latter since U is relatively D-projective.

We get a consequence on homomorphisms almost just as we did in section 10.

Corollary 4 *If U and U_1 are non-projective indecomposable kG-modules lying in B and V and V_1 are the corresponding kN_1-modules lying in b_1 then*

$$\overline{\mathrm{Hom}}_{kG}(U, U_1) \cong \overline{\mathrm{Hom}}_{kN_1}(V, V_1).$$

Proof We have that $V_1^G \cong U_1 \oplus Q_1$, where Q_1 is projective and $U_{N_1} \cong V \oplus W$, as in the theorem. Hence

$$\overline{\mathrm{Hom}}_{kG}(U, U_1) \cong \overline{\mathrm{Hom}}_{kG}(U, U_1 \oplus Q_1)$$
$$\cong \overline{\mathrm{Hom}}_{kG}(U, V_1^G)$$
$$\cong \overline{\mathrm{Hom}}_{kN_1}(U_{N_1}, V_1),$$

where the last isomorphism is exactly as in section 10. Hence,

$$\overline{\mathrm{Hom}}_{kG}(U, U_1) \cong \overline{\mathrm{Hom}}_{kN_1}(V \oplus W, V_1) \cong \overline{\mathrm{Hom}}_{kN_1}(V, V_1)$$

since $\overline{\mathrm{Hom}}_{kN_1}(W, V_1) = 0$ by the description of W.

The structure of B will be derived, using Theorem 3 and Corollary 4, in a number of steps: we shall prove that B has exactly e simple modules, where e is the inertial index of B; the extension of these simple modules will be analyzed; we shall then show that B is a 'Brauer graph algebra with many multiplicities'; finally, Theorem 1 is proved.

Before proceeding, we wish to make a diversion. We have been studying modular representations, that is, kG-modules where k has prime characteristic p. However, the theory of blocks with cyclic defect groups has enormous applications to representation theory in characteristic zero. We want to give the reader some idea of the beauty and power of these results. In general, if K is an algebraically closed field of characteristic zero then the theory of KG-modules is tied up with the theory of kG-modules; we shall describe the consequences for the cyclic theory of KG-modules.

Since K has characteristic zero, the number of simple KG-modules equals the number r of conjugacy classes of elements of G. Let $S_1, \ldots, S_r$ be these simple modules with $S_1 = K$, the trivial one-dimensional KG-module. The character χ_i of S_i is the function from G to K whose value on the group element g is the trace of the linear transformation that g induces on S_i. Since similar linear transformations have equal traces, it follows that χ_i is constant on conjugacy classes. Hence, if $g_1 = 1, g_2, \ldots, g_r$ are representatives of the conjugacy classes of G then the $r \times r$ matrix $(\chi_i(g_j))$, called the *character table* of G, exhibits all the values of all the characters. For example, if G is the symmetric group of degree three then the table is as follows:

	(1)	(2)	(123)
χ_1	1	1	1
χ_2	1	-1	1
χ_3	2	0	-1

This matrix is intimately related to the structure of G and there are numerous magical applications, in which properties of the table lead to theorems on group structure. We shall describe the consequences of the cyclic theory for this remarkable matrix.

Let us first suppose that G has a Sylow p-subgroup P of order p so each block of G either has defect 0 and is a matrix algebra or has P as a defect group and is a Brauer tree algebra. Assume, without loss, that the representatives $g_1 = 1, g_2, \ldots, g_s$ are of order not divisible by p while $g_{s+1}, \ldots, g_r$ have order divisible by p. We wish to break up the characters in a similar way. The *degree* of the character χ_i is the dimension of the module S_i; since the unit element of G induces the identity linear transformation on S_i, its trace is this dimension, that is, $\chi_i(g_1)$, as $g_1 = 1$, is the degree of χ_i and the first column of the character table (the column under g_1) is a list of these degrees. We can suppose that $\chi_1, \ldots, \chi_t$ have degrees not divisible by p while $\chi_{t+1}, \ldots, \chi_r$ do have degrees divisible by p so the character table breaks up

into four submatrices as follows

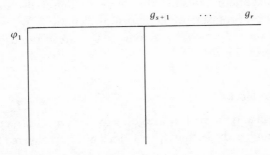

The first result, which is a consequence of the fact that a block of defect zero is a matrix algebra, is that the lower right-hand submatrix has all its entries zero.

To go on we must bring in the group $N = N(P)$ as the next result shows how 'global' information, about G, is locally describable. We wish to look at the character table of N. Since g_i, $i > s$, is of order divisible by p, it is conjugate, by Sylow's theorem and since P has order p, with an element of $N(P)$, so we may assume $g_i \in N(P)$. Then, again as a consequence of Sylow's theorem, the elements $g_{s+1}, \ldots, g_r$ are representatives of the conjugacy classes of elements of $N(P)$ whose order is divisible by p. Since P is normal and abelian in $N(P)$, all the degrees of the characters of $N(P)$ have order not divisible by p, so we can picture the character table of $N(P)$ as divided into two halves, as follows:

The truly fantastic consequence of the cyclic theory is that, with a suitable ordering of characters of $N(P)$, the right-hand half of the character table of

$N(P)$ is the same matrix as the upper right-hand submatrix of the character table of G, except that some of the rows from $N(P)$ must be multiplied by -1 to have this equality. This result depends on deep results and is not easily obtained by using just characters. It leads easily to remarkably restrictive information in the degrees of $\chi_1, \ldots, \chi_t$ and is a powerful tool with many uses.

In general, suppose that P is allowed to be arbitrary, but that B is a block of G with cyclic defect group D, just as in our fixed notation. There is a set of characters of G corresponding to B, also called a block of characters or a block, and a block of characters of N_1 corresponding with b_1 such that there is a coincidence of values, except for some signs, between the two blocks of characters, on the elements of G whose p-part is conjugate with a non-identity element of D and the same elements in N_1.

There are other coincidences of values yet to be put into a general framework and other, more subtle, connections between character values to be understood so there are great discoveries to be made. The structure of blocks with non-cyclic defect groups and the values of characters are the main problem of local representation theory. There are great triumphs ahead.

Exercises

1 Show directly that e divides $p^n - 1$.
2 Prove that the trees given for $SL(2, p)$ are correct.
3 Do the same for the trees for groups with cyclic normal Sylow p-subgroup.
4 Let A be a Brauer tree algebra and let the simple A-modules $T_1, \ldots, T_t$ label the edges emanating from a vertex v and let them be in circular order as given. If v has multiplicity m (taking $m = 1$ if v is not the exceptional vertex) and $q \leqslant mt$ show that there is a uniserial A-module, unique up to isomorphism, of composition length q whose first composition factor is T_1, whose next one is T_2, and so on, using the circular ordering of the T_is.

18 Nilpotent blocks

In this section and the next we are going to analyze the structure of the block b_1. We begin here by letting β_1 be a block of $C(D_1)$ covered by b_1.

Theorem 1 *The block β_1 is a Brauer tree algebra for the Brauer tree with one edge and multiplicity $p^n - 1$.*

In the last section, while discussing the example of the group algebra of a cyclic p-group, the argument given really showed that an algebra was a Brauer tree algebra for the tree with one edge and multiplicity m, if, and only if, the algebra had exactly one simple module and the corresponding indecomposable projective module was uniserial of composition length $m + 1$. Thus, in the case of Theorem 1, we must prove that β_1 has just one simple module, that the corresponding indecomposable projective module is uniserial and, finally, that the composition length of this projective module is p^n. Hence, β_1 will have structure very similar to that of kD. This is an example of what is called a nilpotent block, a concept that can be defined using the synthesis between local group theory and block theory that was developed in section 16.

We begin our analysis by observing that the inertial index of the block β_1 is one. Indeed, by Theorem 15.1, we may assume, without any loss of generality, that β_1 has defect group D and so the inertial index of β_1 divides $|N(D) \cap C(D_1):C(D)|$ and is also not divisible by p. However, $N(D) \cap C(D_1)/C(D)$ is isomorphic with a group of automorphisms of D which are trivial on D_1. But D is a cyclic group of order p^n and so its automorphism group is the direct product of a cyclic group of order $p - 1$ which acts faithfully on the subgroup D_1 of order p and a group of order p^{n-1} which acts trivially on D_1. Thus, the index $|N(D) \cap C(D_1):C(D)|$ is a power of p and our claim is valid. Similarly, if $\bar{\beta}_1$ is the block of $C(D_1)/D_1$ corresponding to β_1 by Lemma 15.6, then $\bar{\beta}_1$ has inertial index one. In fact, $\bar{\beta}_1$ has D/D_1 as a defect group, the normalizer of D/D_1 in $C(D_1)/D_1$ is $N(D) \cap C(D_1)/D_1$ and this induces a p-group of automorphisms of D/D_1.

We shall now assume that we are proving Theorem 17.1 by induction on the order of G; there is no loss of generality in so doing as the whole chapter is a proof of that result. Thus, if D/D_1 is a non-identity group then induction applies to the block $\bar{\beta}_1$ of $C(D_1)/D_1$ and it is a Brauer tree algebra for a tree with one edge, since its inertial index is one. Thus, $\bar{\beta}_1$ has a unique simple module in this case. On the other hand, if $D = D_1$ then $\bar{\beta}_1$ is a block of defect zero and it also has just one simple module. But the simple modules lying in β_1 are just the simple modules lying in $\bar{\beta}_1$, so we have proved the following assertion.

Lemma 2 *There is a unique simple module lying in β_1.*

It is not necessary to use the above induction. Instead, one can prove, again by induction, the special case of Theorem 17.1 for blocks of inertial index one. This requires the arguments of this section and parts of the arguments of later sections. This seems too distracting just for the sake of a lessened induction hypothesis.

The block β_1 and, in fact, any block with a cyclic defect group, has only a finite number of isomorphism classes of indecomposable modules; any algebra with this property is said to be of *finite type*. Indeed, any indecomposable module lying in β_1 has a vertex Q contained in D and a source which is an indecomposable kQ-module and therefore one of finitely many kQ-modules, as Q is a cyclic p-group. Hence, every indecomposable module lying in β_1 is a summand of one of a finite number of modules induced from subgroups of D to $C(D_1)$.

Proposition 3 *If the algebra A is of finite type and has just one simple module then A is a Brauer tree algebra or a matrix algebra.*

Proof Let S be a simple A-module, let d be the dimension of S and let P be the corresponding indecomposable projective A-module. If $\text{rad}(P) = 0$ then $\text{rad} A = 0$, since $A \cong P \oplus \cdots \oplus P$ (d times) as A-modules, so A is semisimple and we are done. Next, suppose that $\text{rad}(P)/\text{rad}^2(P) \cong S$. Let $R = \text{rad}(P)$ so our assumption is that $R/\text{rad}(R) \cong S$ and therefore, by Lemma 5.5, R is a homomorphic image of P. Hence, $\text{rad}(R)/\text{rad}^2(R)$ is a homomorphic image of $\text{rad}(P)/\text{rad}^2(P)$ and so either is isomorphic with S or is zero. But $\text{rad}(R)/\text{rad}^2(R) = \text{rad}^2(P)/\text{rad}^3(P)$ so, continuing in this way, we deduce that each of the successive quotients of the radical series of P is isomorphic with S. Hence P is uniserial and A is a Brauer tree algebra for the tree with one edge and multiplicity one less than the composition length of P. Thus, the only remaining possibility to consider is that $\text{rad}(P)/\text{rad}^2(P) \cong S \oplus \cdots \oplus S$ (r times) with $r > 1$.

Let $\bar{A} = A/\text{rad}^2 A$ and $Q = P/\text{rad}^2(P)$. Since $(\text{rad } A)^2 P = \text{rad}^2(P)$ we have, as $\bar{A}$-modules,

$$\bar{A} \cong Q \oplus \cdots \oplus Q$$

(d times). Thus, S is the only simple $\bar{A}$-module, Q is the corresponding indecomposable projective $\bar{A}$-module and $\text{rad}(Q) \cong S \oplus \cdots \oplus S$ (r times). Since every $\bar{A}$-module is an A-module it suffices to prove that $\bar{A}$ has infinitely many indecomposable modules since this will give the desired contradiction.

We now set $E = \text{End}_{\bar{A}}(Q)$ and shall analyze E. Express $\text{rad}(Q) = S_1 \oplus \cdots \oplus S_r$, as a direct sum, where each S_i is isomorphic with S. Let φ_i, $1 \leqslant i \leqslant r$, be an endomorphism of Q which has kernel $\text{rad}(Q)$ and image S_i and let I be the identity automorphism of Q. We claim that $I, \varphi_1, \ldots, \varphi_r$, is a basis for E. But Q is indecomposable so E is a local algebra and, in particular, $\text{rad } E$ is of codimension one in E while rad E consists of the nilpotent endomorphisms of E. Since $I \notin$ rad E while clearly $\varphi_i^2 = 0$, it suffices to prove that $\varphi_1, \ldots, \varphi_r$ are linearly independent and rad E has dimension r. If $\lambda_i \in k$, $1 \le i \le r$, $\varphi = \sum \lambda_i \varphi_i$ and $x \in Q$ then $\varphi(x) = \sum \lambda_i \varphi_i(x)$. But $\varphi_i(x) \in S_i$ and $\varphi_i(x) \ne 0$ if $x \notin$ rad Q so if $\lambda_j \ne 0$ then $\varphi \ne 0$. Thus, $\varphi_1, \ldots, \varphi_r$ are linearly independent. On the other hand, rad E consists of the endomorphisms with proper images, that is, images contained in rad Q. Hence,

$$\dim \operatorname{rad} E = \dim \operatorname{Hom}_{\bar{A}}(Q, \operatorname{rad} Q)$$
$$= \dim \operatorname{Hom}_{\bar{A}}(S, S_1 + \cdots + S_r)$$
$$= r$$

and our claim is valid.

We shall now construct some E-modules. Since $\varphi_i \varphi_j = 0$, for all i and j, the space spanned by $\varphi_3, \ldots, \varphi_r$ is an ideal of E and is of codimension three. Hence, if V is a vector space over k, X and Y are linear transformations on V such that $X^2 = XY = YX = Y^2 = 0$ then we can make V into an E-module by setting, for $v \in V$, $i > 2$,

$$\varphi_1 v = X(v), \quad \varphi_2 v = Y(v), \quad \varphi_i v = 0, \quad Iv = v.$$

However, in section 4, we constructed, for each positive integer n, a $2n$-dimensional vector space V_n and maps X and Y with these properties. The algebra of linear transformations commuting with X and Y is therefore the endomorphism algebra of the E-module V_n. But we proved, in section 4, that this algebra is local. Hence, E has indecomposable modules of arbitrarily large dimension.

Now $\bar{A} \cong \operatorname{End}_{\bar{A}}(\bar{A})^0$ and $\operatorname{End}_{\bar{A}}(\bar{A}) \cong M_d(E)$ since $\bar{A} \cong Q \oplus \cdots \oplus Q$ (d times). Since E is visibly commutative, we have $\bar{A} \cong M_d(E)$. Hence, in order to complete the proof, we need only show that $M_d(E)$ has indecomposable modules of arbitrarily large dimension. However, if U is an E-module, let U_d consist of column vectors of length d with the d entries from U; thus, in an obvious and natural way, U_d is a module for $M_d(E)$. We claim that the endomorphism algebra of the $M_d(E)$-module U_d is isomorphic with the endomorphism algebra of the E-module U; in particular, if U is indecomposable then so is U_d. Thus, once we establish the claim just made, the $M_d(E)$-modules $(V_n)_d$ will be indecomposable and of arbitrarily high dimension.

As a vector space over k, U_d is isomorphic with the vector space $U \oplus \cdots \oplus U$ (d times) so, by Lemma 2.3 applied to the algebra k, the algebra M of all d by d matrices, with entries linear transformations of U, is isomorphic with the algebra of all linear transformations of U_d: an element of M

obviously defined a corresponding linear transformation of U_d, by the multiplication of a matrix and a column vector, and this correspondence gives the isomorphism. Hence, letting $\bar{e}$ be the linear transformation induced on U by the element e of E, it follows that the endomorphism algebra of the $M_d(E)$-module U_d is isomorphic with the algebra C of all elements of M which commute with each element of $M_d(\bar{E})$, the algebra of all matrices of the form $(\bar{e}_{ij})$ for $e \in E$. In particular, since E has a unit element, each element of C must commute with each of the matrices I_{ij} which have the i,j entry, the identity linear transformation and all other entries zero. This readily implies that each element of C is a 'scalar' matrix, that is, has a constant diagonal entry, one of the linear transformations of U, and all off-diagonal entries zero. But each element of C must commute with each 'scalar' matrix σ_e, which has diagonal entry $\bar{e}$ for $e \in E$. Hence, each element of C is a 'scalar' matrix with diagonal entry from $\mathrm{End}_E(U)$. But all such matrices are in C as it is easy to see that $M_d(\bar{E})$ is generated by all the I_{ij} and σ_e matrices. Hence, our last claim is finally established and the proposition is proved.

Thus, in order to prove Theorem 1, we need only show that the indecomposable projective β_1-module has composition length p^n. But the indecomposable projective $\bar{\beta}_1$-module has composition length p^{n-1}, either by induction, in the case that $n > 1$ or because $\bar{\beta}_1$ has defect zero when $n = 1$. Hence, we need some information about projective modules of a group in terms of projective modules of a quotient group. We shall establish a stronger result than we actually need.

Proposition 4 *Let Q be a normal p-subgroup of G and set $\bar{G} = G/Q$. If S is a simple kG-module let P_S be the corresponding indecomposable projective kG-module and, considering S as a $k\bar{G}$-module, let $\bar{P}_S$ be the corresponding indecomposable projective $k\bar{G}$-module. There is a series of submodules*

$$P_S = P_S^0 \supset P_S^1 \supset \cdots \supset P_S^r = 0$$

so that

$$P_S^i/P_S^{i+1} \cong \bar{P}_S \otimes (\mathrm{rad}^i(kQ)/\mathrm{rad}^{i+1}(kQ)),$$

where kQ is considered as a kG-module using conjugation, $1 \leqslant i \leqslant r$ and r is the radical length of the algebra kQ.

The group G acts on Q by conjugation so kQ can thus be considered as a kG-module so the successive quotients of the radical series of kQ are indeed kG-modules. Before proving this result, suppose that Q is in the center of G.

Thus, each of the modules tensored with $\bar{P}_S$ has G acting trivially so, in particular, the composition length of P_S equals the product of the composition length of $\bar{P}_S$ and the dimension of kQ, that is, the order of Q. Hence, once we prove this proposition then Theorem 1 will be established in full.

Proof We consider P_S as a kQ-module and look at its radical series: set $P_S^i = R^i P_S$ where $R = \mathrm{rad}\ kQ$. Since $g R^i g^{-1} = R^i$, for any g in G, we have that each P_S^i is a kG-submodule. Moreover, if $x \in Q$ then $x - 1 \in R$, by Exercise 3.2, so $(x-1)P_S^i \subseteq P_S^{i+1}$ and P_S^i/P_S^{i+1} is a $k\bar{G}$-module. Conversely, if V is a kG-submodule of P_S^i and P_S^i/V is a $k\bar{G}$-module, that is, Q acts trivially on it, then $(x-1)P_S^i \subseteq V$ for any $x \in Q$. But R is spanned by all such elements $x-1$ so $P_S^{i+1} = R P_S^i \subseteq V$. Thus, P_S^{i+1} is the smallest submodule of P_S^i whose quotient is a $k\bar{G}$-module.

In particular, as $P_S/\mathrm{rad}(P_S) \cong S$ is a $k\bar{G}$-module, we have $P_S^0 = P_S \supseteq \mathrm{rad}(P_S) \supseteq P_S^1$. Hence, P_S^0/P_S^1 is a $k\bar{G}$-module, the quotient of its radical is isomorphic with S and therefore it is also indecomposable. We claim that $P_S^0/P_S^1 \cong \bar{P}_S$; it clearly suffices to prove that P_S^0/P_S^1 is a projective $k\bar{G}$-module. Let φ be a homomorphism of the $k\bar{G}$-module V onto the $k\bar{G}$-module W and let ψ be a homomorphism of P_S^0/P_S^1 to W. Let η be the natural homomorphism of P_S onto P_S^0/P_S^1 so $\psi\eta$ is a homomorphism of P_S to W. Since P_S is a projective kG-module, there is a kG-homomorphism σ of P_S to V such that $\varphi\sigma = \psi\eta$, that is, the following diagram is commutative:

$$
\begin{array}{ccc}
P_S & \xrightarrow{\eta} & P_S^0/P_S^1 \\
\sigma \downarrow & & \downarrow \psi \\
V & \xrightarrow{\varphi} & W
\end{array}
$$

Hence, in order to show that P_S^0/P_S^1 is projective, we need only prove that there is a homomorphism ρ of P_S^0/P_S^1 to V such that $\rho\eta = \sigma$, as then $\varphi\rho\eta = \varphi\sigma = \psi\eta$ which implies that $\varphi\rho = \psi$. However, η is the natural map, so we need only see that P_S^1 is in the kernel of σ; but V is a $k\bar{G}$-module so this is certainly the case.

We claim now that

$$P_S^i/P_S^{i+1} \cong P_S^0/P_S^1 \otimes R^i/R^{i+1},$$

where R and all its powers are kG-modules using conjugation. Since $P_S^0/P_S^1 \cong \bar{P}_S$ we will have proved the proposition once we have established this claim. However, if $r \in R^j$ and $v \in P_S^i$ then $rv \in P_S^{i+t}$. Therefore there is a linear transformation of $(P_S^0/P_S^1) \otimes (R^i/R^{i+1})$ to P_S^i/P_S^{i+1} which sends

$$(u + P_S^1) \otimes (r + R^{i+1})$$

to $ru + P_S^{i+1}$, whenever $r \in R^i$ and $u \in P_S^0$. The image is all of P_S^i/P_S^{i+1} since $RP_S^i = P_S^{i+1}$. Moreover, this linear transformation is a kG-homomorphism. In fact, if $g \in G$ then

$$g((u + P_S^1) \otimes (r + R^{i+1})) = (gu + P_S^1) \otimes (grg^{-1} + R^{i+1})$$

is mapped to $grg^{-1}gu + P_S^{i+1}$ which is $g(ru + P_S^{i+1})$. We have now constructed a kG-homomorphism of $P_S^0/P_S^1 \otimes R_i/R_{i+1}$ onto P_S^i/P_S^{i+1}. In particular, the dimension of the right-hand module is at most the dimension of the left-hand one so if we sum over all i we get that $|Q| \dim \bar{P}_S \geqslant \dim P_S$, as $|Q| = \dim kQ$. Moreover, if this last inequality is an equality then $P_S^0/P_S^1 \otimes R^i/R^{i+1}$ and P_S^i/P_S^{i+1} will have the same dimension, for each i, and therefore, be isomorphic and the proposition will be proved. However,

$$|G| = \sum_S (\dim S) \dim P_S$$

$$\leqslant \sum_S (\dim S)|Q| \dim \bar{P}_S$$

$$= |Q||\bar{G}| = |G|$$

so we are finished.

Exercises

1   With the notation of the proof of Proposition 3, if U and V are E-modules then $\operatorname{Hom}(U, V) \cong \operatorname{Hom}(U_d, V_d)$, where we mean E-homomorphisms and $M_d(E)$-homomorphisms, respectively.

2   (cont.) If W is an $M_d(E)$-module then there is an E-module V such that $W \cong V_d$.

3   With the notation of Proposition 4, if S and T are simple kG-modules then $\dim \operatorname{Hom}_{kG}(P_S, P_T) = |Q| \dim \operatorname{Hom}_{kG}(\bar{P}_S, \bar{P}_T)$.

4   Use Proposition 4 to give an alternative treatment of the structure of projective modules for groups with a normal cyclic Sylow p-subgroup.

5   If B has a unique simple module then B is a Brauer tree algebra for a tree with one edge.

19 Local case

In this section we shall obtain the structure of b_1, as described in Theorem 17.2, which we now restate.

Theorem 1 *The block b_1 is a Brauer tree algebra for a star with e edges and exceptional vertex in the center with multiplicity $(p^n - 1)/e$.*

We shall use the block β_1 from the previous section to obtain the structure of b_1. Denote $C_1 = C(D_1)$ so β_1 is a block of C_1 which is a normal subgroup of $N_1 = N(D_1)$ and we let I_1 be the stabilizer of β_1 in N_1 in the sense of section 15. The proof of Theorem 1 falls into two parts, the first of which is the proof of the next result.

Proposition 2 *The block $\beta_1^{I_1}$ is a Brauer tree algebra for a star with e edges and exceptional vertex in the center with multiplicity $(p^n - 1)/e$.*

The first step in establishing Proposition 2 is as follows.

Lemma 3 *The index $|I_1 : C_1|$ equals e.*

Set $N = N(D)$ and $C = C(D)$. Let γ be the Brauer correspondent of the block β_1 of the group C_1: thus, γ is a block of $N \cap C_1 = N_{C_1}(D)$ with a defect group D and $\gamma^{C_1} = \beta_1$. Hence, from section 15 we know that if β is a block of C covered by γ then β has a defect group D and $\beta^{N_1 \cap C} = \gamma$. Hence, $\beta^G = B$, since $\gamma^{C_1} = \beta_1$ and $\beta_1^G = G$, so if I is the stabilizer of β in N then the definition of the inertial index gives us that $|I : C| = e$. To prove the lemma we shall show that $I/C \cong I_1/C_1$. We shall do this using the First Isomorphism Theorem by proving that $I_1 \cap C = C_1$ and $IC_1 = I_1$. However, $N \cap C_1/C$ is a p-group, as it is isomorphic with a group of automorphisms of D which are trivial on D_1, while I/C has order not divisible by p by Lemma 15.5. Hence, $N \cap C_1/C \cap I/C = 1$, that is, $N \cap C_1 \cap I = C$. But $I \subseteq N$ so $I \cap C_1 = C$.

Finally, suppose that $x \in I_1$; we must prove that $x \in IC_1$. But x stabilizes β_1 so $x^{-1}Dx$ is also a defect group of β_1 and therefore there is z in C_1 with $x^{-1}Dx = z^{-1}Dx$. Hence, if $y = xz^{-1}$ then $y \in N \cap I_1$ and $x = yz$ so we need only show that $y \in IC_1$.

The blocks of C covered by γ are β and its conjugates in $N \cap C_1$. Since $I \cap (N \cap C_1) = C$, the number of these conjugates is $|N \cap C_1 : C| = q$ which must be a power of p since $N \cap C_1/C$ is isomorphic with a group of automorphisms of D which are trivial on D_1. Now y stabilizes γ since γ is the Brauer correspondent of β_1 while y stabilizes β_1, as $y \in I_1$, and y normalizes D, as $y \in N$. Therefore, the cyclic group $\langle y \rangle$ generated by y permutes by conjugation the q blocks of C covered by γ. But I is normal in N since N/I is abelian: it is isomorphic with a group of automorphisms of D and all automorphisms of D are power maps. Hence, I is the stabilizer in N of each conjugate of β in N and so, in particular, all the orbits of $\langle y \rangle$ on the conjugates of β have the same size. Therefore, y^q stabilizes all the q blocks of C covered by γ so certainly $y^q \in I$. But $|I_1 : C_1|$ is not divisible by p, as I_1/C_1 is isomorphic with a group of automorphisms of D_1, so yC_1 and y^qC_1 generate the same subgroup of I_1/C_1 as q is a power of p. Hence, $y \in \langle y^q \rangle C_1$, that is, $y \in IC_1$, and the proposition is established.

We shall now start investigating the structure of $\beta_1^{I_1}$ by determining the simple modules lying in that block.

Lemma 4 *If S is a simple β_1-module then S has exactly e extensions to a kI_1-module and these are the simple modules in $\beta_1^{I_1}$.*

Thus, there are exactly e isomorphism classes of kI_1-modules whose restriction to C_1 is isomorphic with S and these are exactly the simple kI_1-modules which lie in the block $\beta_1^{I_1}$.

Proof For each $c \in C_1$ let $\sigma(c)$ be the linear transformation induced on S by c. Let $x \in I_1$ be chosen so that xC_1 is a generator of the cyclic group I_1/C_1. Since I_1/C_1 has order e, $y = x^e \in C_1$. In order to show there is an extension of S to I_1 it suffices to find a linear transformation t of S such that

$$t^e = \sigma(y),$$
$$t\sigma(e)t^{-1} = \sigma(t\sigma t^{-1})$$

for all $c \in C_1$, inasmuch as it is then clear how to define the action of any element $t^i c, c \in C_1$ on S and that this defines a module. However, S is stable under the action of I_1: the conjugate of S by any element of I_1 is another simple module lying in β_1 and so is isomorphic with S. But $x \in I_1$ so there is a linear transformation u of S such that

$$u\sigma(c)u^{-1} = \sigma(xcx^{-1})$$

for all $c \in C_1$. Therefore

$$u^e\sigma(c)u^{-e} = \sigma(x^ecx^{-e}) = \sigma(y)\sigma(c)\sigma(y)^{-1}$$

and so $\sigma(y)^{-1}u^e$ is an endomorphism of the kC_1-module S. Hence, $\sigma(y)^{-1}u^e = \lambda 1_S$, where $0 \neq \lambda \in k$ and 1_S is the identity linear transformation of S. Choose $\mu \in k$ such that $\mu^{-e} = \lambda$ and set $t = \mu u$ and this clearly has the desired properties.

Hence, let T be an extension of S to I_1 so T is a kI_1-module with $T_{C_1} = S$. In particular, T is certainly simple. Let us also consider the induced module S^{I_1}. We have

$$S^{I_1} = (T_{C_1})^{I_1} \cong (T_{C_1} \otimes k)^{I_1} \cong T \otimes k^{I_1}.$$

But I_1/C_1 is a cyclic group of order e where e is not divisible by p. Therefore, the free $k[I_1/C_1]$-module k^{I_1} is the direct sum of e distinct one-dimensional modules so S^{I_1} is the direct sum of e extensions of S to I_1 (not necessarily distinct). But

$$\mathrm{Hom}_{kI_1}(T, S^{I_1}) \cong \mathrm{Hom}_{kC_1}(T_{C_1}, S) \cong \mathrm{Hom}_{kC_1}(S, S) \cong k,$$

so exactly one of these e summands is isomorphic with T. But T was any extension so the e summands must be distinct and any extension is isomorphic with exactly one of them.

It remains therefore only to establish the second part of the assertion of the lemma. If T is any extension of S to I_1 then $T_{C_1} = S$ so the block of I_1 in which T lies must cover β_1. But $C_1 = C(D_1) \supseteq C(D)$ so $\beta_1^{I_1}$ is the only block of I_1 which covers β_1. Hence, the e extensions of S all lie in $\beta_1^{I_1}$. On the other hand, suppose that U is a simple module lying in $\beta_1^{I_1}$. Hence, U_{C_1} is the direct sum of modules, one from each conjugate of β_1 in I_1. But β_1 is invariant under I_1 so U_{C_1} lies in β_1. However, U_{C_1} is semisimple, by Clifford's theorem, and S is the only simple module lying in β_1 so $U_{C_1} \cong S \oplus \cdots \oplus S$ (r times). But

$$\mathrm{Hom}_{kC_1}(U_{C_1}, S) \cong \mathrm{Hom}_{kI_1}(U, S^{I_1})$$

and S^{I_1} is the direct sum of e distinct simple modules. Hence, $r = 1$, U is an extension of S and the lemma is proven.

Lemma 5 *If U is a kI_1-module lying in $\beta_1^{I_1}$ then* $\mathrm{rad}(U) = \mathrm{rad}(U_{C_1})$.

Proof If V is a $b_1^{I_1}$-module and V_{C_1} is semisimple then $V_{C_1} \cong S \oplus \cdots \oplus S$, where S is the simple β_1-module, by Lemma 4. But $V | (V_{C_1})^{I_1}$, as C_1 contains the defect group D of $\beta_1^{I_1}$, and S^{I_1} is semisimple by the proof of Lemma 4 so V is semisimple if V_{C_1} is semisimple. However, $\mathrm{rad}(U_{C_1}) = \mathrm{rad}(kC_1)U$ is a kI_1-submodule of U, as C_1 is normal in I_1, and $U/\mathrm{rad}(U_{C_1})$ is a semisimple kC_1-module so $U/\mathrm{rad}(U_{C_1})$ is semisimple as a kI_1-module, that is, $\mathrm{rad}(U) \subseteq \mathrm{rad}(U_{C_1})$. But the reverse inequality holds, by Clifford's theorem, and the lemma is proved.

Now let T be a simple $\beta_1^{I_1}$-module which is an extension of the simple β_1-module S and let P be the indecomposable projective kI_1-module corresponding with T. Thus, $P/\mathrm{rad}(P) \cong T$ so $P_{C_1}/\mathrm{rad}(P_{C_1}) \cong S$, by Lemma 5. However, P_{C_1} is also projective so it is the indecomposable projective kC_1-module corresponding with S. Thus, P_{C_1} is uniserial of composition length p^n with each composition factor isomorphic with S. Hence, by Lemma 5, P is uniserial of composition length p^n with each composition factor one of the extensions of S. In particular, $\mathrm{rad}(P)/\mathrm{rad}^2(P)$ is a simple $\beta_1^{I_1}$-module so, by the proof of Lemma 4, it is of the form $T \otimes W$ where W is a one-dimensional $k[I_1/C_1]$-module. Thus, $V = P/\text{rad}^2(P)$ is a non-split extension of the simple module $T \otimes W$ by the simple module T.

Next, suppose that U is also a simple $\beta_1^{I_1}$-module so, by the proof of Lemma 4, there is a one-dimensional $k[I_1/C_1]$-module L with $U \cong T \otimes L$. Thus, $V \otimes L$ is an extension of the simple module $T \otimes W \otimes L$ by the simple module $T \otimes L$, that is, $U \otimes W$ by U. Moreover, this extension is not split. Indeed, let L_1 be the one-dimensional kC_1-module which is the 'inverse' of L, so the element x of C_1 acts on L_1, by multiplication by $\lambda^{-1} \in k$ whenever x acts on L by multiplication by λ; thus $L \otimes L_1 \cong k$. Hence, $V \otimes L \otimes L_1 \cong V$ so if $V \otimes L$ is a split extension then so also would be $V \cong (V \otimes L) \otimes L_1$. Hence $V \otimes L$ is a non-split extension and is therefore a homomorphic image of the indecomposable projective corresponding with U. Thus, the second composition factor of any indecomposable projective module in $\beta_1^{I_1}$ is always the tensor product of W and the first composition factor.

We can now argue just as in the last paragraph of section 5: the successive factors of the radical series of P are T, $T \otimes W$, $T \otimes W \otimes W$, and so on, and the same rule applies to any indecomposable projective module in $\beta_1^{I_1}$. We also assert that if we set $T_1 = T$, $T_{i+1} = T_i \otimes W$ then $T_1, T_2, \ldots, T_e$ are the distinct simple $\beta_1^{I_1}$-modules and $T_{e+1} \cong T_1$. Indeed, since we can define an 'inverse' W_1 for W, it follows that the map of isomorphism classes of simple $\beta_1^{I_1}$-modules, given by tensoring with W, is a permutation. Our assertion is simply that there is just one orbit under the action of W. If this is not the case then there are distinct simple $\beta_1^{I_1}$-modules $U_1, \ldots, U_f$, $f < e$, with $U_i \otimes W \cong U_{i+1}$, $1 \leqslant i < f$, $U_f \otimes W \cong U_1$ and no other such simple module, when tensored with W, is isomorphic with any U_j, $1 \leqslant j \leqslant f$. Hence, by our description of indecomposable projectives above, the only composition factors of the projectives corresponding with $U_1, \ldots, U_f$ are $U_1, \ldots, U_f$ and no other projective has any of these simple modules as a composition factor. Hence, $U_1, \ldots, U_f$ are the simple modules in a block, which is a contradiction, as $f < e$.

Now consider the tree labelled with $T_1, \ldots, T_e$ as follows:

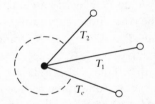

Let the multiplicity be $(p^n - 1)/e$. This shows that $\beta_1^{I_1}$ is a Brauer tree algebra, as claimed in Proposition 2. Indeed, the successive composition factors of the projective indecomposable module corresponding with T_1 are T_1, T_2, ..., where there are p^n factors. This sequence is just the sequence $T_1, \ldots, T_e$ repeated $(p^n - 1)/e$ times followed by T_1, which is just what is required in the definition of a Brauer tree algebra. Since similar assertions hold for each T_i, $i \geqslant 1$, Proposition 2 is proved.

The results which enable us to describe b_1 using $\beta_1^{I_1}$ are given as follows:

Lemma 6 *If V is a $\beta_1^{I_1}$-module then V^{N_1} is a b_1-module and any b_1-module is isomorphic with such an induced module. Moreover, if V_1 and V_2 are $\beta_1^{I_1}$-modules then induction defines an isomorphism of $\mathrm{Hom}_{kI_1}(V_1, V_2)$ onto $\mathrm{Hom}_{kN_1}(V_1^{N_1}, V_2^{N_2})$.*

Proof Let $g_1 = 1, g_2, \ldots, g_r$ be representatives for the cosets of I_1 in N_1. Set $\beta_i = g_i \beta_i g_i^{-1}$ as $\beta_1, \ldots, \beta_r$ are the distinct conjugates of β_1 in N_1, as I_1 is the stabilizer of β_1. Since V lies in $\beta_1^{I_1}$, we have that V_{C_1} lies in β_1 as β_1 is the only block of C_1 covered by $\beta_1^{I_1}$. Thus,

$$(V^{N_1})_{C_1} \cong (g_1 \otimes V)_{C_1} \oplus \cdots \oplus (g_r \otimes V)_{C_1}$$

and $(g_i \otimes V)_{C_1}$ lies in β_i since $(g_i \otimes V)_{C_1}$ is the conjugate of V_{C_i} by g_i. Hence, each indecomposable summand of V^{N_1} lies in a block covering $\beta_1, \ldots, \beta_r$. But $D \subseteq C_1$ and $\beta_1^{I_1} = b_1$ so only b_1 covers these blocks and thus V^{N_1} lies in b_1.

On the other hand, suppose W is a kN_1-module lying in b_1. Thus,

$$W_{C_1} = \beta_1 W + \cdots + \beta_r W$$

is the decomposition of W_{C_1} with respect to the blocks of C_1. Moreover, since β_1 is stabilized by I_1, $\beta_1 W$ is a kI_1-module. Its restriction to C_1 lies in β_1 so $\beta_1 W$ is a kI_1-module lying in $\beta_1^{I_1}$. Since the β_is are all conjugate in N_1 it follows that all the $\beta_i W$ have the same dimension and that $\beta_1 W$ generates W as a kN_1-module. This immediately implies that $\dim W = r \dim(\beta_1 W)$ so $W \cong (\beta_1 W)^{N_1}$ and the first assertion of the lemma is proved.

Finally, if $\varphi \in \mathrm{Hom}_{kI_1}(V_1, V_2)$ then $\varphi^{N_1} \in \mathrm{Hom}_{kN_1}(V_1^{N_1}, V_2^{N_2})$ and the map which sends φ to φ^{N_1} is a one-to-one linear transformation (since after identifying each V_i with $1 \otimes V_i$ we have that the restriction of φ^{N_1} to V_1 is φ). Hence, to conclude the proof, it suffices to show the two vector spaces in question have the same dimension. However,

$$\mathrm{Hom}_{kN_1}(V_1^{N_1}, V_2^{N_2}) \cong \mathrm{Hom}_{kI_1}((V_1^{N_1})_{I_1}, V_2) \cong \mathrm{Hom}_{kI_1}(\beta_1^{I_1}(V_1^{N_1})_{I_1}, V_2)$$

since V_2 lies in $\beta_1^{I_1}$. However, $(g_i \otimes V)_{C_1}$ lies in β_i so if $i > 1$ then $(g_i \otimes V)_{I_1}$ can have no non-zero summand lying in $\beta_1^{I_1}$ as $\beta_1^{I_1}$ covers only β_1. Hence, $\beta_1^{I_1}(V_1^{N_1})_{I_1} \cong V_1$ and we are done.

We are now in a position to establish Theorem 1. The last result will enable us to see that the properties of b_1-modules are virtually identical with the properties of $\beta_1^{I_1}$-modules so that, in particular, Theorem 1 will be a consequence of Proposition 2. The next lemma details some of the relationships between properties of $\beta_1^{I_1}$-modules and b_1-modules.

Lemma 7 *If U, V, W, V_1, V_2 are $\beta_1^{I_1}$-modules, $\varphi \in \mathrm{Hom}_{kI_1}(V, V)$ and $\psi \in \mathrm{Hom}_{kI_1}(V_1, V_2)$ then the following assertions hold:*

(1) *φ is the identity if, and only if, φ^{N_1} is the identity;*

(2) *ψ is one-to-one (onto) if, and only if, ψ^{N_1} is one-to-one (onto).*

(3) *V is simple (semisimple, projective) if, and only if, V^{N_1} is simple (semisimple, projective);*

(4) *If $\alpha \in \mathrm{Hom}_{kI_1}(U, V)$ and $\lambda \in \mathrm{Hom}_{kI_1}(V, W)$ then the sequence*

$$0 \to U \overset{\alpha}{\to} V \overset{\lambda}{\to} W \to 0$$

is exact if, and only if, the sequence

$$0 \longrightarrow U^{N_1} \overset{\alpha^{N_1}}{\longrightarrow} V^{N_1} \overset{\lambda^{N_1}}{\longrightarrow} W \longrightarrow 0$$

is exact;

(5) $\mathrm{rad}(V^{N_1}) \cong \mathrm{rad}(V)^{N_1}$, $(V/\mathrm{rad}(V))^{N_1} \cong V^{N_1}/\mathrm{rad}(V^{N_1})$.

Proof The first assertion is clear from the construction (but can be proved in a more functorial way – see the exercises) as is the second. Now V is simple if, and only if, every non-zero homomorphism from a $\beta_1^{I_1}$-module to V is onto, so the first part of (3) holds because of (2). If V is semisimple then there are simple modules $S_1, \ldots, S_t$ and homomorphisms $\lambda_i \in \mathrm{Hom}_{kI_1}(S_i, V)$, $\pi_i \in \mathrm{Hom}_{kI_1}(V, S_i)$, $1 \leqslant i \leqslant t$, such that $\pi_i \lambda_i = 1_{S_i}$ and $\sum \lambda_i \pi_i = 1_V$. Thus,

$$\pi_i^{N_1} \lambda_i^{N_1} = 1_{S_i^{N_1}}, \quad \sum \lambda_i^{N_1} \pi_i^{N_1} = 1_{V^{N_1}}$$

so $V^{N_1} \cong S_1^{N_1} \oplus \cdots \oplus S_t^{N_1}$, which is semisimple as each $S_i^{N_1}$ has already been proved simple. Conversely, if V^{N_1} is semi-simple then $V^{N_1} \cong S_1^{N_1} \oplus \cdots \oplus S_t^{N_1}$ for simple $\beta_1^{I_1}$-modules $S_1, \ldots, S_t$, by the first part of (3) and by Lemma 6, and there are maps, of the form $\lambda_i^{N_1}, \pi_i^{N_1}$, by Lemma 6, satisfying the above conditions so the λ_i, π_i also satisfy the above conditions, by Lemma 6 and part (1) of Lemma 7, and so $V \cong S_1 \oplus \cdots \oplus S_t$. Next, suppose that V^{N_1} is projective. Let δ be a homomorphism of U onto W and γ a homomorphism of V to W: we wish to show there is a homomorphism ρ of V to U such that $\delta\rho = \gamma$ in order to establish that V is projective. However, δ^{N_1} is a homomorphism of U^{N_1} onto W^{N_1} and γ^{N_1} is a homomorphism of V^{N_1} to W^{N_1} so, as V^{N_1} is projective, there is a homomorphism, necessarily of the form ρ^{N_1}, where $\rho \in \mathrm{Hom}_{kI_1}(V, U)$, such that $\delta^{N_1}\rho^{N_1} = \gamma^{N_1}$. Hence, $\delta\rho = \gamma$ as $(\delta\rho)^{N_1} = \gamma^{N_1}$. Conversely, if V is projective it is easy to show that V^{N_1} is projective, by a similar argument. The fourth statement is also clear since α is one-to-one if, and only if, α^{N_1} is one-to-one and λ is onto if, and only if, λ^{N_1} is onto while dim V is the sum of dim U and dim W if, and only if, dim V^{N_1} is the sum of dim U^{N_1} and dim W^{N_1}.

Before proving the fifth part of the lemma, notice that part (4) implies that each homomorphic image of V^{N_1} is of the form $(V/U)^{N_1}$, where U is a submodule of V. Indeed, let W^{N_1} be the image of V^{N_1} under the homomorphism λ^{N_1}, let U be the kernel of λ so (4) implies that $W_1^{N_1} \cong (V/U)^{N_1}$ as $(V/U)^{N_1} \cong V^{N_1}/U^{N_1} \cong W^{N_1}$.

Turning to (5), we have, in particular, applying this to $V^{N_1}/\mathrm{rad}(V^{N_1})$, that the composition length of $V^{N_1}/\mathrm{rad}(V^{N_1})$ is at no more than the composition length of $(V/\mathrm{rad}(V))^{N_1}$. But the exact sequence

$$0 \to \mathrm{rad}(V) \to V \to V/\mathrm{rad}(V) \to 0$$

gives the exact sequence

$$0 \to \mathrm{rad}(V)^{N_1} \to V^{N_1} \to (V/\mathrm{rad}(V)^{N_1} \to 0$$

so $\mathrm{rad}(V)^{N_1} \supseteq \mathrm{rad}(V^{N_1})$ inasmuch as $(V/\mathrm{rad}(V))^{N_1}$ is semisimple. Hence, the above inequality on composition lengths implies $\mathrm{rad}(V)^{N_1} = \mathrm{rad}(V^{N_1})$ and that (5) holds.

Let us now prove Theorem 1. Let $T_1, \ldots, T_e$ be the simple $\beta_1^{I_1}$-modules so $T_1^{N_1}, \ldots, T_e^{N_1}$ are the simple b_1-modules. Let $P_1, \ldots, P_e$ be the indecomposable projective modules corresponding with $T_1, \ldots, T_e$. Thus, $P_1^{N_1}, \ldots, P_e^{N_1}$ are projective by part (3) of Lemma 7, $P_i^{N_1}/\mathrm{rad}(P_i^{N_1}) \cong T_i^{N_1}$, by part (5) of Lemma 7, so $P_i^{N_1}$ is the indecomposable projective module corresponding with $T_i^{N_1}$. If $T_1, \ldots, T_e$ are in the circular order of the Brauer tree for $\beta_1^{I_1}$ then the successive quotients of the radical series of P_i are T_i, T_{i+1}, and so on, and therefore, by part (5) of Lemma 7, the successive quotients of the radical series of $P_i^{N_1}$ are $T_i^{N_1}, T_{i+1}^{N_1}$, and so on. Hence, b_1 is a Brauer tree algebra for

the tree

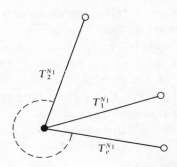

where the multiplicity is $(p^n - 1)/e$.

We conclude the section with one consequence.

Corollary 8 *The block b_1 has exactly ep^n isomorphism classes of indecomposable modules each of which is uniserial.*

This follows from the structure of the projective modules exactly in the same way as the analogous statement was proved for groups with a normal cyclic Sylow p-subgroup.

Exercises

1 Let N be a normal subgroup of G and S a simple kN-module stable under conjugation by G.
  (i) If G/N is a p-group then S has a unique extension to G.
  (ii) If G/N is cyclic of order $p^f e$, where $p \nmid f$, then S has exactly f extensions to G.

2 With the notation of Lemma 7, show that $\varphi = 1_V$ if, and only if, whenever α is a non-zero homomorphism of a module M to V then $\varphi\alpha = \alpha$. Use this to give another proof of part (1).

3 (cont.) Show that ψ is one-to-one if, and only if, whenever α is a non-zero homomorphism of a module M to V_1 then $\varphi\alpha \neq 0$. Use this to give another proof of the one-to-one part of (2) in Lemma 7 and then give a similar functorial argument for the onto part.

4 (cont.) Give a proof of part (4) of Lemma 7 without referring to dimensions.

20 Projective covers

We have used the fact that every module is a homomorphic image of a projective module and in this section we shall prove some general properties about such homomorphisms. In the case of group algebras this will lead us to the Heller operator, which we shall use extensively in the remainder of the chapter. Although we could have introduced all this material much earlier we have waited until we were able to make the applications of these ideas.

If U and V are A-modules then we have defined $\overline{\operatorname{Hom}}_A(U, V)$, the quotient of $\operatorname{Hom}_A(U, V)$ by the vector space of all homomorphisms of U to V which factor through a projective module. Henceforth, we shall call such a homomorphism a *projective* homomorphism. We now have the problem of calculating $\overline{\operatorname{Hom}}(U, V)$ in order to use it in the way it arises, as explained at the end of section 17. At first, it seems that to determine whether a homomorphism is projective or not, it is necessary to survey all projective modules. However, this is not the case, as we shall now see.

Lemma 1 *Let φ be a homomorphism of the projective A-module Q onto the A-module V. If α is a homomorphism of the A-module to V then α is projective if, and only if, α factors through Q.*

Proof Certainly, if α factors through Q then α is projective since Q is projective. Conversely, suppose that P is a projective A-module and $\gamma \in \operatorname{Hom}_A(U, P)$, $\delta \in \operatorname{Hom}_A(P, V)$ with $\alpha = \delta\gamma$ so α is projective. Since φ maps Q onto V and P is projective there is $\rho \in \operatorname{Hom}_A(P, Q)$ with $\varphi\rho = \delta$. Hence, $\rho\gamma \in \operatorname{Hom}_A(U, Q)$ and $\varphi(\rho\gamma) = \delta\gamma = \alpha$ so α factors through Q.

Hence, in order to test whether α is projective we need only look at one projective module, one that can be mapped onto V. Next, we shall see that there is a canonical choice available in selecting Q. If Q is a projective A-module then we say that Q is a *projective cover* of V if $Q/\operatorname{rad} Q \cong V/\operatorname{rad} V$. Thus, $V/\operatorname{rad} V$ is a homomorphic image of Q so, as Q is projective, there is a homomorphism of Q to V whose image W has the property that $W + \operatorname{rad} V = V$. But each maximal submodule of V contains $\operatorname{rad} V$ so this means that W cannot be a proper submodule, that is, $W = V$ and V is a homomorphic image of Q. The next result explains the uniqueness of Q and the homomorphism of Q onto V.

Lemma 2 *Let Q be a projective cover of the A-module V and let φ be a homomorphism of Q onto V. Let ψ be a homomorphism of the projective module P onto V.*

(1) *There is a homomorphism ρ of P onto Q such that $\varphi\rho = \psi$.*

(2) *There is a direct decomposition $P = R + K$ where K is the kernel of ρ and R is a projective cover of V.*

Since $\rho(P) = Q$ and K is the kernel of ρ we must now have that ρ maps R isomorphically onto Q so R is also a projective cover of V. Moreover, $\psi(K) = 0$, as $\psi = \varphi\rho$ and $\rho(K) = 0$, so the restriction of ψ to R must map R onto V. Thus, P is the direct sum of a projective cover of V and another projective module K and ψ is trivially determined by a homomorphism of R onto V.

If P is a projective cover of V then $K = 0$. Indeed, if $K \neq 0$ then

$$P/\text{rad } P \cong R/\text{rad } R \oplus K/\text{rad } K$$

$$\cong Q/\text{rad } Q \oplus K/\text{rad } K$$

$$\cong V/\text{rad } V \oplus K/\text{rad } K$$

so $P/\text{rad } P \not\cong V/\text{rad}(V)$. Hence, we have a commutative diagram

$$\begin{array}{ccc} P & \xrightarrow{\psi} & V \\ \rho\downarrow & & \downarrow 1 \\ Q & \xrightarrow{\varphi} & V \end{array}$$

and ρ is an isomorphism of P onto Q so we have a uniqueness not just for Q but for Q together with the homomorphism φ of Q onto V.

Proof Since P is projective and φ maps Q onto V there is a homomorphism ρ of P to Q such that $\varphi\rho = \psi$. We claim that $\rho(P) = Q$. Indeed, suppose that $\rho(P)$ is a proper submodule of Q so it is contained in a maximal submodule of Q; hence, $\rho(P) + \text{rad}(Q)$ is also a proper submodule of Q. But $Q/\text{rad}(Q) \cong V/\text{rad}(V)$, by hypothesis, and φ maps Q onto V so φ must induce an isomorphism of $Q/\text{rad}(Q)$ onto $V/\text{rad}(V)$. In particular, $\varphi(\rho(P) + \text{rad}(Q))$ is a proper submodule of Q, so $Q = \varphi(\rho(P))$ is also a contradiction.

If K is the kernel of ρ then P/K is projective, as $P/K \cong Q$, so there is a submodule R of P such that $R + K = P$ is a direct sum. Thus, the restriction of ρ to R is an isomorphism onto Q so R is also a projective cover of V and (2) is proved.

With these results in mind, we select, for each A-module V, a projective cover PV and a homomorphism π_V of PV onto V. We denote the kernel of π_V by ΩV. Hence, ΩV is determined up to isomorphism by V and if ψ is any homomorphism of a projective module P onto V then the kernel of ψ is isomorphic with $\Omega V \oplus K$, where K is a projective module.

At this point we shall specialize to the case of the group algebra kG for the remainder of the section. Our first task is to dualize the notions we have introduced. An injective (that is, projective) kG-module Q is an *injective envelope* of the kG-module V if $soc(V) \cong soc(Q)$ in which case there is a one-to-one homomorphism of V to Q which then must map $soc(V)$ isomorphically onto $soc(Q)$. There is the dual version of Lemma 2 which is easy to state and prove and this is left to the reader. In particular, the injective envelope of a kG-module, as well as the embedding, is canonical in the obvious sense. For each kG-module V let IV be an injective envelope and let λ_V be a one-to-one homomorphism of V into IV. We set $\Omega^{-1}V = IV/\lambda_V(V)$ so $\Omega^{-1}V$ is determined up to isomorphism by V and whenever λ is a one-to-one homomorphism of V into an injective kG-module Q then $Q/\lambda(V) \cong \Omega^{-1}(V) \oplus R$ where R is an injective module. We can also say that a homomorphism α of a kG-module U to a kG-module V is *injective* if it factors through an injective module; thus, α is injective if, and only if, it is projective. Moreover, just as in the first lemma, if φ is a one-to-one homomorphism of U into the injective module Q then α is injective if, and only if, α factors through Q.

Here is a useful criterion which is an application of these ideas.

Lemma 3 *Let α be a projective homomorphism of the kG-module U to the kG-module V. If α is onto (respectively, one-to-one) then U (respectively, V) has a non-zero projective summand.*

This is very useful when U (or V) has no non-zero projective summand, that is, is a *projective-free* module.

Proof First, suppose that α is onto. Since α is projective there is a homomorphism ρ of U to PV such that $\pi_V \rho = \alpha$, that is, the following diagram is commutative:

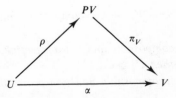

But π_V induces an isomorphism of $PV/\mathrm{rad}(PV)$ onto $V/\mathrm{rad}(V)$ and $\alpha(U) = V$ so we must have $PV = \rho(U) + \mathrm{rad}(PV)$. As before, this implies that $\rho(U) = PV$ so U has a direct summand isomorphic with PV.

Similarly, if α is one-to-one we have a commutative diagram as follows:

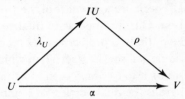

But λ_U maps soc(U) one-to-one onto soc(IU) and α is one-to-one so ρ must map soc(IU) into V. Hence, ρ has zero kernel (or its socle is in soc(IU)) and V has a summand isomorphic with IU.

In general, if U is any kG-module then U can be written as the direct sum of a projective-free module and a projective module. Fix such a decomposition and let $\Omega^0 U$ be the projective-free summand so $\Omega^0 U$ is determined by U, up to isomorphism.

Lemma 4 *If* U *is a* kG-*module, then*

$$\Omega\Omega^{-1}U \cong \Omega^0 U \cong \Omega^{-1}\Omega U.$$

Proof Let $V=\Omega^0 U$ so $U=V+Q$ is a direct sum for a suitable projective submodule Q. Since Q is its own projective cover, we have $PU \cong PV \oplus Q$ so $\Omega U \cong \Omega V$. Since we similarly have $\Omega^{-1}U \cong \Omega^{-1}V$, it is sufficient to show that $\Omega\Omega^{-1}V \cong V \cong \Omega\Omega^{-1}V$.

But $IV/V = \Omega^{-1}V$ so IV is a projective module and $\Omega^{-1}V$ is a homomorphic image with corresponding kernel V. Thus, $V \cong \Omega\Omega^{-1}V \oplus R$ for a suitable projective module R. But $V=\Omega^0 V$ is projective-free so $R=0$ and $V \cong \Omega\Omega^{-1}V$. Similarly, we also have that $V \cong \Omega^{-1}\Omega V$.

Note that in the course of the proof we saw that if V is a projective-free kG-module then IV is a projective cover of $\Omega^{-1}V$ and similarly PV is an injective envelope of ΩV.

Theorem 5 *Let* U_1, U_2, V *be indecomposable and non-projective* kG-*modules.*

(1) ΩV *is also indecomposable and non-projective.*

(2) *If* $\Omega U_1 \cong \Omega U_2$ *then* $U_1 \cong U_2$.

(3) *There is an indecomposable and non-projective* kG-*module* U *with* $\Omega U \cong V$.

Hence, the Heller 'operator' Ω defines a one-to-one correspondence on the isomorphism classes of indecomposable and non-projective kG-modules.

Proof First, ΩV is projective-free: by Lemma 4,

$$\Omega^0 \Omega V \cong \Omega \Omega^{-1}(\Omega V) = \Omega(\Omega^{-1} V) \cong \Omega \Omega^0 V \cong \Omega V.$$

Suppose that ΩV is decomposable so $\Omega V = W_1 + W_2$ is a direct sum for suitable projective-free modules W_1 and W_2. Hence,

$$V = \Omega^0 V \cong \Omega^{-1} \Omega V \cong \Omega^{-1} W_1 \oplus \Omega^{-1} W_2$$

is a non-zero direct sum, which is a contradiction, and (1) is proved.

If $\Omega U_1 = \Omega U_2$ then we have

$$U_1 = \Omega^0 U_1 \cong \Omega^{-1} \Omega U_1 \cong \Omega^{-1} \Omega U_2 \cong \Omega^0 U_2 = U_2.$$

Finally, $\Omega^{-1} V$ is also indecomposable, by a similar argument, so $V = \Omega^0 V \cong \Omega \Omega^{-1} V$ and $U = \Omega^{-1} V$ gives the third assertion.

The Heller operator also preserves homomorphisms, at least modulo projective ones.

Lemma 6 *If U and V are kG-modules then*

$$\overline{\mathrm{Hom}}_{kG}(U, V) \cong \overline{\mathrm{Hom}}_{kG}(\Omega U, \Omega V).$$

Proof We may assume, without loss of generality, that U and V are projective-free. Indeed, $U = \Omega^0 U + Q$, $V = \Omega^0 V + R$ are direct sums with Q and R projective so $\overline{\mathrm{Hom}}_{kG}(U, V) \cong \overline{\mathrm{Hom}}_{kG}(\Omega^0 U, \Omega^0 V)$ while $\Omega U \cong \Omega \Omega^0 U$ and $\Omega V \cong \Omega \Omega^0 V$.

If $\alpha \in \mathrm{Hom}_{kG}(U, V)$ then there is a commutative diagram (not necessarily unique) as follows:

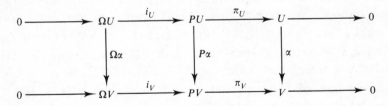

Here i_U and i_V are the embeddings, as ΩU and ΩV are submodules of PU and PV, respectively, and $P\alpha$ and $\Omega\alpha$ are suitable homomorphisms. Indeed, π_V maps PV onto V and $\alpha\pi_U$ is a homomorphism of the projective module PU to V so there is a homomorphism $P\alpha$ such that $\pi_V P\alpha = \alpha\pi_U$. It follows that $P\alpha(\Omega U) \subseteq \Omega V$: $\pi_V P\alpha(\Omega U) = \alpha\pi_U(\Omega U) = \alpha(0) = 0$. Hence, $\Omega\alpha$ exists to complete the diagram as claimed. Conversely, any homomorphism of ΩV to ΩV arises in this way for a suitable α and consequent choice of $P\alpha$: since U and V are projective-free, PU and PV are injective envelopes of ΩU and ΩV

and U and V are isomorphic with $\Omega^{-1}\Omega U$ and $\Omega^{-1}\Omega V$ so a similar and dual argument applies.

Suppose that α is projective; we claim that $\Omega\alpha$, chosen as above, is also projective. In fact, there is $\delta \in \operatorname{Hom}_{kG}(U, PV)$ such that $\pi_V\delta = \alpha$. It follows that $\delta\pi_U - P\alpha$ maps PU not just to PV but into ΩV: $\pi_V(\delta\pi_U - P\alpha) = (\pi_V\delta)\pi_U - \pi_V P\alpha = \alpha\pi_U - \pi_V P\alpha = 0$. Moreover,

$$(P\alpha - \delta\pi_U)i_U = (P\alpha)i_U = i_V\Omega\alpha$$

so $\gamma = P\alpha - \delta\pi_U$ regarded as a map of PU to ΩV satisfies $\gamma i_U = \Omega\alpha$ and $\Omega\alpha$ is projective as claimed.

Conversely, if $\Omega\alpha$ is projective then so is α. Indeed, there is $\gamma \in \operatorname{Hom}_{kG}(\Omega U, \Omega V)$ with $\gamma i_U = \Omega\alpha$ since PU is an injective envelope of ΩU and α is injective. Now $(P\alpha - i_V\gamma)i_U = (P\alpha)i_U - i_V\Omega\alpha = 0$ so ΩU is in the kernel of $P\alpha - i_V\gamma \in \operatorname{Hom}_{kG}(PU, PV)$. Hence, there is, by the fundamental theorem on homomorphisms, $\delta \in \operatorname{Hom}_{kG}(U, PV)$ such that $P\alpha - i_V\gamma = \delta\pi_U$. We assert that $\pi_V\delta = \alpha$ so α is projective as claimed. It suffices to prove that $\pi_V\delta\pi_U = \alpha\pi_U$ as π_U maps PU onto U. However,

$$\pi_V\delta\pi_U = \pi_V P\alpha - \pi_V i_V\mu$$
$$= \pi_V P\alpha$$
$$= \alpha\pi_U.$$

If $\alpha' \in \operatorname{Hom}_{kG}(U, V)$ and $P\alpha'$ and $\Omega\alpha'$ are chosen suitably then $P\alpha - P\alpha'$ and $\Omega\alpha - \Omega\alpha'$ provide a commutative diagram for $\alpha - \alpha'$. In particular, if $\alpha = \alpha'$ then $\alpha - \alpha' = 0$ is certainly projective so $\Omega\alpha - \Omega\alpha'$ is also. Hence, while $\Omega\alpha$ depends on α and a choice of $P\alpha$, its image in $\overline{\operatorname{Hom}}_{kG}(\Omega U, \Omega V)$ depends just on α. Since the above argument for differences also works for sums and scalar multiples we have a linear transformation of the vector space $\operatorname{Hom}_{kG}(U, V)$ to $\overline{\operatorname{Hom}}_{kG}(\Omega U, \Omega V)$. Since this sends a projective map to zero, we have a linear transformation of $\overline{\operatorname{Hom}}_{kG}(U, V)$ to $\overline{\operatorname{Hom}}_{kG}(\Omega U, \Omega V)$. Similarly, starting with an element of $\operatorname{Hom}_{kG}(\Omega U, \Omega V)$, we get a linear transformation of $\overline{\operatorname{Hom}}_{kG}(\Omega U, \Omega V)$ to $\overline{\operatorname{Hom}}_{kG}(U, V)$. But α and $\Omega\alpha$ determine each other, up to projective homomorphisms, so these two linear transformations are inverse to each other and the lemma is proved.

Our final result relates the Heller operator and the Green correspondence.

Proposition 7 *Let U be a non-projective indecomposable kG-module.*

(1) *The modules U and ΩU have the same vertices.*

(2) *If V is a Green correspondent of U then ΩV is a Green correspondent of ΩU.*

Proof Suppose that H is a subgroup of G and U is relatively H-projective. Let W be a kH-module such that $U \mid W^G$. The exact sequence

$$0 \to \Omega W \to PW \to W \to 0$$

yields the exact sequence

$$0 \to (\Omega W)^G \to (PW)^G \to W^G \to 0.$$

Now $W^G \cong U \oplus X$, for a kG-module X, $(PW)^G$ is a projective module and so $(PW)^G \cong PU \oplus PX \oplus R$, when R is a projective module. Moreover, we have a commutative diagram, with vertical arrows isomorphisms

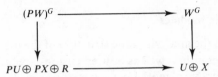

where the homomorphism at the bottom maps PU onto U, PX onto X and R to 0. Hence, ΩU is a summand of the kernel $(\Omega W)^G$, that is, ΩU is relatively H-projective. A similar statement of course also holds for Ω^{-1}, so if ΩU is relatively H-projective then so is $\Omega^{-1}\Omega U \cong U$. This proves the first assertion.

Let R be a vertex of U, Q a p-subgroup of G containing R and H a subgroup of G containing $N(Q)$ so that the Green correspondence between G and H applies to modules with vertex R and that V is then the Green correspondent of U. Since we now know that ΩU and ΩV are indecomposable and have vertex R, it suffices to prove that $\Omega V \mid \Omega U$. However, the exact sequence

$$0 \to \Omega U \to PU \to U \to 0$$

yields the exact sequence

$$0 \to (\Omega U)_H \to (PU)_H \to U_H \to 0.$$

But $(PU)_H$ is projective so we can proceed just as in the preceding paragraph and so complete the proof.

Exercises

1 Let U and V be kG-modules.
   (a) $\Omega(U \oplus V) \cong \Omega U \oplus \Omega V$.
   (b) $\Omega^{-1}U \cong (\Omega U^*)^*$.
   (c) $\Omega(U \otimes V) \cong \Omega^0(\Omega U \otimes V) \cong \Omega^0(U \otimes \Omega V)$.
   (d) $\Delta(kG) \otimes U \cong \Omega U \oplus Q$, where Q is projective.

(e) If n is a positive integer let $\Omega^n U = \Omega \cdots \Omega U$ (n times) and $\Omega^{-n} U = \Omega^{-1} \cdots \Omega^{-1} U$ (n times). For any integers r and s,

$$\Omega^r \Omega^s U \cong \Omega^{r+s} U.$$

2 (cont.) The vector space $\overline{\mathrm{Hom}}_{kG}(U, V)$ is isomorphic with the vector space of fixed points for G in the kG-module $\Omega^0(U^* \otimes V)$. Use this to deduce another proof of Lemma 6.

3 (cont.) If U is indecomposable and non-projective and S is a source for U then ΩS is a source for ΩU.

21 Simple modules

We begin the analysis of the structure of the block B by describing the simple modules in B as well as how extensions are built from them. The block b_1 is a Brauer tree algebra for a star so the e simple modules in b_1 have a canonical circular ordering. Let $V_0, \ldots, V_{e-1}$ be these e modules in the canonical order. For any integer i, let V_i be the simple b_1-module such that $V_i = V_t$ where $i \equiv t$ (modulo e) and $0 \leqslant t < e$. If j is an integer, $1 \leqslant j \leqslant q = p^n$, let V_{ij} be the indecomposable b_1-module with $V_{ij}/\mathrm{rad}(V_{ij}) \cong V_i$ and V_{ij} of composition length j. Thus, V_{ij} is uniserial with composition factors, in order, $V_i, V_{i+1}, \ldots, V_{i+j-1}$. Moreover, the eq modules V_{ij} are the indecomposable b_1-modules up to isomorphism.

Theorem 1 *The block B has exactly e simple modules. Each simple b_1-module is a quotient of the Green correspondent of exactly one of the simple B-modules and is a submodule of the Green correspondent of exactly one of the simple B-modules.*

Proof Let S_1 and S_2 be distinct simple B-modules with Green correspondents T_1 and T_2, that is, the corresponding b_1-modules. Suppose that the simple b_1-module V_i is a homomorphic image of T_1 and of T_2. Thus, as T_1 and T_2 are non-projective indecomposable modules and non-isomorphic, we have $T_1 \cong V_{is}$, $T_2 \cong V_{it}$ where $1 \leqslant s, t < q$ and $s \neq t$. Therefore, after renumbering if necessary, we may assume that $s < t$ so V_{is} is a homomorphic image of V_{it}, being isomorphic with the unique quotient module of V_{it} that has composition length s. Thus, $\mathrm{Hom}_{kN_1}(T_2, T_1) \neq 0$ and, by Lemma 20.3, $\overline{\mathrm{Hom}}_{kN_1}(T_2, T_1) \neq 0$. Hence, $\overline{\mathrm{Hom}}_{kG}(S_2, S_1) \neq 0$ so certainly $\mathrm{Hom}_{kG}(S_2, S_1) \neq 0$, a contradiction. Thus, the simple modules $T_1/\mathrm{rad}(T_1)$ and $T_2/\mathrm{rad}(T_2)$ are non-isomorphic.

Moreover, since the e indecomposable projective b_1-modules are uniserial, their eq submodules are pairwise non-isomorphic as they are determined by their socles and their composition length. Hence, if two indecomposable b_1-modules have isomorphic socles then one is isomorphic with a submodule of the other. Hence, by a similar (but dual) argument to the above one, we also have that the simple modules $\mathrm{soc}(T_1)$ and $\mathrm{soc}(T_2)$ are not isomorphic.

On the other hand, let W_i be the Green correspondent of the simple b_1-module V_i so W_i is an indecomposable B-module. There is a simple B-module S such that $\mathrm{Hom}_{kG}(S, W_i) \neq 0$. Hence, by Lemma 20.3, $\overline{\mathrm{Hom}}_{kG}(S, W_i) \neq 0$, and so $\overline{\mathrm{Hom}}_{kN_1}(T, V_i) \neq 0$ where T is the Green correspondent of S. Thus, certainly $\mathrm{Hom}_{kN_1}(T, V_i) \neq 0$ so we have shown that each simple b_1-module is a homomorphic image of the Green correspondent of a simple B-module. Hence, the map which corresponds $T/\mathrm{rad}(T)$ to S is a one-to-one correspondence. In particular, the first two assertions of the theorem are established. If we now consider, using the above notation, a simple module which is a homomorphic image of W_i we similarly have that each simple b_1-module is isomorphic with the socle of the Green correspondent of a simple B-module and the last assertion of the theorem also holds.

Let us establish some notation in the light of the theorem. Let $S_0, S_1, \ldots, S_{e-1}$ be the simple B-modules chosen so that $T_i/\mathrm{rad}(T_i) \cong V_i$, where T_i is the Green correspondent of S_i, $0 \leqslant i < e$. Hence, there is a permutation π of the set $\{0, 1, \ldots, e-1\}$ such that $V_{\pi(i)} \cong \mathrm{soc}(T_i), 0 \leqslant i < e$. For any b_1-module V we let $l(V)$ be its composition length.

Lemma 2 $\overline{\mathrm{Hom}}_{kN_1}(T_i, T_j)$ *is of dimension zero or one according as $i \neq j$ or $i = j$.*

Proof We have $\overline{\mathrm{Hom}}_{kN_1}(T_i, T_j) \cong \overline{\mathrm{Hom}}_{kG}(S_i, S_j)$ and $\overline{\mathrm{Hom}}_{kG}(S_i, S_j) = \mathrm{Hom}_{kG}(S_i, S_j)$ by Lemma 20.3.

Lemma 3 *If $0 \neq \theta \in \mathrm{Hom}_{kN_1}(V_{is}, V_{jt})$ then θ is projective if, and only if, $l(\theta(V_{is})) \leqslant s + t - q$.*

Proof Let $Y = \theta(V_{is})$ and set $r = l(Y)$. Since V_{jq} is projective and $V_{jq}/\mathrm{rad}(V_{jq}) \cong V_j$ we have that V_{jq} is a projective cover of V_{jt}; let φ be a homomorphism of V_{jq} onto V_{jt} and let X be the pre-image of Y. Now $l(V_{jt}/Y) = l(V_{jt}) - l(Y) = t - r$ so $l(V_{jq}/X) = t - r$, as $V_{jq}/X \cong V_{jt}/Y$, and therefore $l(X) = q - t + r$.

Suppose that θ is projective. There is a homomorphism ρ of V_{is} to V_{jq} such that $\varphi\rho = \theta$. In particular, $\rho(V_{is}) \subseteq X$. But X is uniserial, being a submodule of V_{jq}, and $Y \neq 0$ so each proper submodule of X is mapped by φ to a proper submodule of Y. Hence $\rho(V_{is}) = X$ and $l(V_{is}) \geqslant l(X)$, that is, $s \geqslant q - t + r$ as claimed.

On the other hand, suppose that $s \geqslant q - t + r$. We shall prove, by induction on r, that θ is projective. Now Y is a non-zero homomorphic image of the uniserial module V_{is} so $Y/\mathrm{rad}(Y) \cong V_i$. Hence, $X/\mathrm{rad}(X) \cong V_i$ as X is uniserial and $\varphi(X) = Y$. Recapitulating, we have that X is uniserial, $X/\mathrm{rad}(X) \cong V_i$ and $l(X) = q - t + r$. But V_{is} is uniserial, $V_{is}/\mathrm{rad}(V_{is}) \cong V_i$ and $l(V_{is}) = s \geqslant q - t + r$. Hence, V_{is} has a quotient module which is uniserial, has radical quotient isomorphic with V_i and has composition length $q - t + r$, so this quotient is isomorphic with X. Hence, there is a homomorphism σ of V_{is} onto X so $\varphi\sigma(V_{is}) = Y = \theta(V_{is})$. Since V_{is} is uniserial, $\varphi\sigma$ and θ have the same kernel, so, by the fundamental theorem on homomorphisms, there is an automorphism α of Y such that $\alpha\varphi\sigma = \theta$. But $\mathrm{End}(Y)$ is local so $\alpha = \lambda 1_Y + \eta$, where η is a nilpotent endomorphism of Y. In particular, $\eta(Y)$ is a proper submodule of Y so $(\lambda\varphi\sigma - \theta)V_{is}$ is a proper submodule of Y. If it is zero then $\lambda\varphi\sigma = \theta$ and we are done, while if it is non-zero then, by induction, $\lambda\varphi\sigma - \theta$ is projective so θ is, too.

Before proceeding, we wish to record how the Heller operator applies to b_1-modules. As we have just seen, V_{jq} is a projective cover of V_{jt}. Thus, ΩV_{jt} is isomorphic with the kernel of a homomorphism of V_{jq} onto V_{jt}; in particular, $l(\Omega V_{jt}) = q - t$. The first t composition factors in the composition series of V_{jt} are $V_j, V_{j+1}, \ldots, V_{j+t-1}$. Hence, the $(t+1)$th such factor, namely V_{j+t}, is the radical quotient of the kernel of a homomorphism of V_{jq} onto V_{jt}. Thus, ΩV_{jt} has radical quotient V_{j+t} and $l(\Omega V_{jt}) = q - t$ so $\Omega V_{jt} \cong V_{j+t, q-t}$.

For convenience's sake, we shall say that an indecomposable b_1-module V is *short* if $l(V) \leqslant e$ and is *tall* if $l(V) \geqslant q - e$.

Lemma 4 *Each module T_i is short or tall.*

Proof Of course, if $q \leqslant 2e$ then all the indecomposable b_1-modules are short or tall, so we shall assume that $q > 2e$. First, suppose that $e < l(T_i) \leqslant q/2$. The composition factors of V_i are thus $V_i, V_{i+1}, \ldots, V_{i+e}, \ldots$ and, since $V_{i+e} \cong V_i$, it follows that T_i has a proper submodule with radical quotient isomorphic with V_i. Hence, V_i has an endomorphism with a proper image and dim $\mathrm{Hom}_{kN_1}(T_i, T_i) \geqslant 2$. However, by Lemma 3,

$$\mathrm{Hom}_{kN_1}(T_i, T_i) \cong \overline{\mathrm{Hom}}_{kN_1}(T_i, T_i) \cong \overline{\mathrm{Hom}}_{kG}(S_i, S_i)$$

and this is a contradiction.

Next, suppose that $q/2 \leqslant l(T_i) < q - e$. Hence, $e < l(\Omega T_i) \leqslant q/2$, as $l(T_i) + l(\Omega T_i) = q$, so we deduce, just as in the previous paragraph, that $\overline{\mathrm{Hom}}_{kN_1}(\Omega T_i, \Omega T_i) \cong \mathrm{Hom}_{kN_1}(\Omega T_i, \Omega T_i)$ has dimension at least two. But, by Lemma 20.6,

$$\overline{\mathrm{Hom}}_{kN_1}(\Omega T_i, \Omega T_i) \cong \overline{\mathrm{Hom}}_{kN_1}(T_i, T_i) \cong \overline{\mathrm{Hom}}_{kG}(S_i, S_i),$$

so we again have a contradiction.

The next result is now clearly applicable to the modules T_i.

Lemma 5 *If V is an indecomposable b_1-module which is short or tall and W is an indecomposable b_1-module then each of the vector spaces*

$$\overline{\mathrm{Hom}}_{kN_1}(V, W), \quad \overline{\mathrm{Hom}}_{kN_1}(W, V), \quad \overline{\mathrm{Hom}}_{kN_1}(\Omega V, W), \quad \overline{\mathrm{Hom}}_{kN_1}(\Omega V, W)$$

has dimension at most one.

Proof First, suppose that V is short and let $V/\mathrm{rad}(V) \cong V_i$. If W has no composition factor isomorphic with V_i then certainly $\mathrm{Hom}_{kN_1}(V, W) = 0$. If W has composition factors isomorphic with V_i then we claim that W has a unique submodule Y such that $Y/\mathrm{rad}(Y) \cong V_i$ and Y is short. Indeed, if Y is the submodule of W such that $Y/\mathrm{rad}(Y)$ is the last composition factor in the composition series of W isomorphic with V_i then $l(Y) < e$: the composition factors of Y, in order, must be $V_i, V_{i+1}, \ldots, V_{i+f}$ and we must have $f < e$, as $V_{i+e} \cong V_i$, so $l(Y) = f < e$. On the other hand, if Z is a submodule of W with $Z/\mathrm{rad}(Z) \cong V_i$ and $Z \neq Y$ then Z properly contains Y and $l(Z/Y) \geqslant e$ since each simple b_1-module will be a composition factor of Z/Y. Thus $l(Z) > e$ and Z is not short. Thus our claim about Y holds. Hence, since any non-zero homomorphic image of V is short and has radical quotient isomorphic with V_i, if φ and θ are non-zero homomorphisms of V to W then $\varphi(V) = Y = \theta(V)$. Thus, reasoning just as in the proof of Lemma 3, there is $0 \neq \lambda \in k$ such that $\varphi - \lambda\theta$ has image properly contained in Y. But $\varphi - \lambda\theta$ is also homomorphism of V to W so $\varphi - \lambda\theta = 0$ and we have proved that $\mathrm{Hom}_{kN_1}(V, W)$ is of dimension at most one.

We can similarly show that $\mathrm{Hom}_{kN_1}(W, V)$ is of dimension at most one. If $W/\mathrm{rad}(W) \cong V_j$ then either V_j is not a composition factor of V and $\mathrm{Hom}_{kN_1}(W, W) = 0$ or W has a unique submodule X such that $X/\mathrm{rad}(X) \cong$

V_j, in which case X is the image of any non-zero homomorphism of W to V and $\mathrm{Hom}_{kN_1}(W, V)$ is of dimension at most one, just as above.

Now $\overline{\mathrm{Hom}}_{kN_1}(\Omega V, W) = 0$ if W is projective while $\overline{\mathrm{Hom}}_{kN_1}(\Omega V, W) \cong \overline{\mathrm{Hom}}_{kN_1}(V, \Omega^{-1} W)$ by Lemma 20.6 if W is not projective. But $\Omega^{-1} W$ is also indecomposable so $\overline{\mathrm{Hom}}(\Omega V, W)$ is of dimension at most one. Similarly, we can deal with $\overline{\mathrm{Hom}}_{kN_1}(W, \Omega V)$ using $\Omega^{-1} W$ and our previous results.

This last result allows us to establish an important general fact about B-modules.

Proposition 6 *If U is an indecomposable B-module then* $\mathrm{soc}(U)$ *and* $U/\mathrm{rad}(U)$ *are each multiplicity-free.*

That is, $\mathrm{soc}(U)$ and $U/\mathrm{rad}(U)$ are each the direct sum of simple submodules, no two of which are isomorphic.

Proof Let S_j be one of the simple B-modules; it suffices to prove that $\mathrm{Hom}_{kG}(S_j, U)$ and $\mathrm{Hom}_{kG}(U, S_j)$ are each of dimension at most one as these vector spaces are isomorphic with $\mathrm{Hom}_{kG}(S_j, \mathrm{soc}(U))$ and $\mathrm{Hom}_{kG}(U/\mathrm{rad}(U), S_j)$, respectively. However, if U is projective then this is clear, while if U is not projective and V is its Green correspondent then, by Lemma 20.3,

$$\mathrm{Hom}_{kG}(S_j, U) \cong \overline{\mathrm{Hom}}_{kG}(S_j, U) \cong \overline{\mathrm{Hom}}_{kG}(T_j, V)$$

so $\mathrm{Hom}_{kG}(S^j, U)$ is as claimed by Lemma 5. Similarly, $\mathrm{Hom}_{kG}(U, S_j)$ is of dimension at most one and the result is proved.

We are now ready to consider extensions of simple B-modules.

Proposition 7 *If U is a non-projective indecomposable B-module and* $0 \leqslant j < e$ *then there is, up to isomorphism, at most one non-split exact sequence.*

$$0 \to S_j \to M \to U \to 0$$

and this exists if, and only if,

$$\overline{\mathrm{Hom}}_{kN_1}(\Omega V, T_j) \neq 0$$

where V is the Green correspondent of U.

By the uniqueness we mean that if

$$0 \to S_j \to M' \to U \to 0$$

is another such sequence then there is an isomorphism θ of M onto M' and an automorphism φ of S_j such that the following diagram commutes:

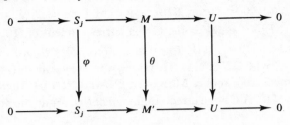

Hence, $M \cong M'$ in a way compatible with the way they are constructed.

Proof First, suppose we have a sequence as in the theorem. Thus, M has a submodule S with $S \cong S_j$ and $M/S \cong U$ while S is not a direct summand of M. It follows that $\text{rad}(M) \supseteq S$: otherwise $\text{rad}(M) \cap S = 0$, $S + \text{rad}(M)/\text{rad}(M)$ is a direct summand of $M/\text{rad}(M)$ and S is a direct summand of M. We now claim that we have a commutative diagram

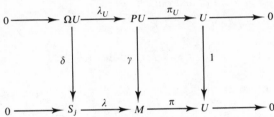

where the lower sequence is the given one and δ and γ are onto. Indeed, since PU is projective there is a homomorphism γ such that $\pi\gamma = \pi_U$. Since $S = \lambda(S_j) \subseteq \text{rad}(M)$ we have that π induces an isomorphism of $M/\text{rad}(M)$ onto $U/\text{rad}(U)$, so, as in previous arguments, $\gamma(PU) = M$. We also have, as before, that δ exists with $\gamma\lambda_U = \lambda\delta$ and $\delta(\Omega U) = S_j$. But $\Omega U/\text{rad}(\Omega U)$ has a unique summand isomorphic with S_j, by Proposition 6, so ΩU has a unique maximal submodule X with $\Omega U/X \cong S_j$; hence X is the kernel of δ and we now have a commutative diagram

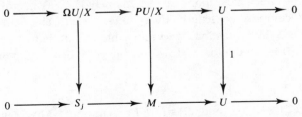

where the vertical maps are isomorphisms and the first row is exact. This immediately implies the uniqueness we claimed. Moreover, we now have $\mathrm{Hom}_{kG}(\Omega U, S_j) \neq 0$ so $\overline{\mathrm{Hom}_{kG}(\Omega U, S_j)} \neq 0$, by Lemma 20.3, and therefore $\overline{\mathrm{Hom}_{kN_1}(\Omega V, T_j)} \neq 0$ as ΩV is the Green correspondent of ΩU.

Conversely, if $\overline{\mathrm{Hom}_{kN_1}(\Omega V, T_j)} \neq 0$ then $\mathrm{Hom}_{kG}(\Omega U, S_j) \neq 0$ and ΩU has a submodule X with $\Omega U/X \cong S_j$. Therefore, we have a commutative diagram just as immediately above. Moreover, $PU/\mathrm{rad}(PU) \cong U/\mathrm{rad}(U)$, so $X \subseteq \mathrm{rad}(PU)$ and $\Omega U/X$ is not a direct summand of PU/X and the proposition is proved.

We shall now apply this result to give detailed information.

Proposition 8 *Let U be an indecomposable and non-projective B-module with Green correspondent V_{is}. If $0 \leqslant j < e$ then there is a non-split exact sequence*
$$0 \to S_j \to M \to U \to 0$$
with M indecomposable if, and only if, either
$$s + l(T_j) > q \quad and \quad i = \pi(j)$$
or
$$s + l(T_j) \leqslant q \quad and \quad i + s \equiv j \ (modulo\ e).$$

Before proving this result let us record a direct consequence.

Corollary 9 *If $0 \leqslant i, j < e$ then there is a non-split exact sequence*
$$0 \to S_j \to M \to S_i \to 0$$
if, and only if, either
$$l(T_i) + l(T_j) > q \quad and \quad i = \pi(j)$$
or
$$l(T_i) + l(T_j) \leqslant q \quad and \quad j \equiv \pi(i) + 1 \ (modulo\ e).$$

In fact, this is obtained by letting $U = S_i$ so the Green correspondence V becomes the Green correspondent of S_i, namely $V_{i,l(T_i)}$, as $T_i/\mathrm{rad}(T_i) \cong V_i$ and $l(T_i)$ is its composition length. Everything is clear, since any non-split extension of S_j by S_i must be indecomposable, except that we need to show $i + l(T_i) \equiv \pi(i) + 1$ (modulo e). However, the composition factors of T_i, in order, are
$$V_i, V_{i+1}, \ldots, V_{i+l(T_i)-1}$$
and so $\mathrm{soc}(T_i) \cong V_{i+l(T_i)-1}$ and $\pi(i) \equiv i + l(T_i) - 1$ (modulo e) by definition of the permutation π. Let us turn to the proof of the proposition.

Proof We shall first explain why Corollary 10.2 carries over to this situation so that, in particular, there is a non-split extension of S_j by U if, and only if, there is a non-split extension of T_j by V_{is}. Indeed, in section 10 we studied the case of trivial intersections while Theorem 17.3 shows we have almost the same situation when dealing with B-modules and b_1-modules. The only difference is that in Theorem 17.3 the module W need not be projective but may contain summands lying in blocks of N_1 other than b_1. However, this causes no problem when carrying over the arguments on extensions from Corollary 10.2. For example, if U is a kN_1-module, V is a submodule, U/V lies in b_1 and V lies in a block of N_1 other than b_1 then V is a direct summand of U: the identity element of b_1 annihilates V and induces the identity on U/V so U is the direct sum of V and the product of that identity element with U.

The indecomposability of M has further consequences because of Theorem 17.3 and the proof of Corollary 10.2. If M is not projective then M_{N_1} is the direct sum of a module L, a projective module and modules lying in blocks other than b_1, where there is a non-split exact sequence

$$0 \to T_j \to L \to V_{is} \to 0$$

and L is the direct sum of a non-projective indecomposable module and a projective module (possibly zero). If M is projective then the same situation occurs, except that now L is projective.

In this latter case, we have $l(T_j) + s = q$, since $l(T_j) + s < 2q$ and $l(T_j) + s \equiv 0$ (modulo q). Since V_i is a homomorphic image of L, we deduce that $L \cong V_{iq}$. But then $V_{j,l(T_j)}$ is the submodule of V_{iq} of composition length $l(T_j)$ so $j \equiv i + s$ (modulo e).

Next, suppose that M is not projective so neither is L. If $l(L) = s + l(T_j) \leqslant q$ then L can have no projective summand and so it is indecomposable and $s + l(T_j) < q$. We may argue as in the preceding paragraph, so $j \equiv i + s$ (modulo e) and $L \cong V_{i,l(T_j)+s}$. If $l(L) = s + l(T_j) > q$ then L is the direct sum of an indecomposable module and a (non-zero) projective module. Therefore, we must have $L/\mathrm{rad}\ L \cong V_i \oplus V_j$, since $T_j/\mathrm{rad}(T_j) \cong V_j$ and $V_{is}/\mathrm{rad}\ V_{is} \cong V_i$, so the projective summand is isomorphic with $V_{i,q}$ or $V_{j,q}$. We assert that it is $V_{i,q}$; assume the opposite so, in particular, $i \neq j$. Since $T_j \cong V_{j,l(T_j)}$ we have

$$0 \to V_{j,l(T_j)} \to L \to V_{i,s} \to 0$$

so L has a submodule isomorphic with $V_{j,l(T_j)}$ and the corresponding quotient module has its radical quotient isomorphic with V_i and not V_j. But the projection of the submodule isomorphic with $V_{j,l(T_j)}$ to the projective module isomorphic with $V_{j,q}$, in a decomposition of L, cannot be onto, as $V_{j,l(T_j)}$ is indecomposable and not projective, so the quotient of L by this submodule has V_j as a homomorphic image, a contradiction. We next claim that $i = \pi(j)$. We already have

$$L \cong V_{i,q} \oplus V_{j,s+l(T_j)-q}$$

so $\mathrm{soc}(L) \cong V_i \oplus V_{j+s+l(T_j)-q-1}$. We need only prove that the submodule of L isomorphic with T_j contains a submodule isomorphic with V_i, inasmuch as $\mathrm{soc}(T_j) \cong V_{\pi(j)}$. But, if this is not the case, then the projection of this submodule to the second summand in the above decomposition of L must be one-to-one and so the quotient of L, by this submodule isomorphic with T_j, has composition length at least q, which is a contradiction. Therefore, we have proved all the implications of the proposition in one direction.

We shall now establish that there is a non-split exact sequence

$$0 \to T_j \to L \to V_{is} \to 0$$

where L is, respectively, non-projective and indecomposable, projective and indecomposable, the direct sum of a non-projective indecomposable and a projective module, according as, respectively, $s + l(T_j) < q$ and $i + s \equiv j$, $s + l(T_j) = q$ and $i + s \equiv j$, $s + l(T_j) > q$ and $i = \pi(j)$. The discussion at the beginning of the proof shows this suffices to conclude the proof in the first and third cases and that in the second case there is a projective module M and a non-split exact sequence

$$0 \to S_j \to M \to U \to 0$$

where U is the correspondent of V_{is} and so is indecomposable and non-projective. But then U is isomorphic with the direct sum of $\Omega^{-1}(S_j)$ and a projective module with the second summand non-zero only if M is decomposable so M must again be indecomposable.

We proceed by analogy with the proofs of Lemmas 10.4 and 10.5, which were established not only for their use in section 10 but also as preparation for this step. In the first two cases set $L = V_{i,s+l(T_j)}$, in analogy with Lemma 10.4, and L has a submodule isomorphic with T_j and corresponding quotient isomorphic with $V_{i,s}$ by the conditions given. In the third case, mimicking Lemma 10.5, let $L = V_{i,q} \oplus V_{j,s+l(T_j)-q}$. Since $j = \pi(i)$ we have $\mathrm{soc}(T_j) \cong V_i$ so, as $V_{i,q}$ is injective there is a submodule W of V_{iq} isomorphic with $T_j \cong V_{j,l(T_j)}$. Now $l(T_j) > s + l(T_j) - q$ so there is a homomorphism φ of W onto $V_{j,s+l(T_j)-q}$. The submodule of L consisting of all elements $(w, \varphi(w))$, $w \in W$, is isomorphic with T_j. It is not a summand, as L has no summand of composition length $l(T_j)$ with radical quotient V_j. Moreover, the radical quotient of L/W is isomorphic with V_i, as is easily seen by arguing as in Lemma 10.5, so $L/W \cong V_{is}$ as $l(L/W) = s$. Hence, the proposition is proved in full.

Exercises

1 If U and V are b_1-modules and dim $U + $ dim $V \leqslant q$ then $\mathrm{Hom}_{kN_1}(U, V) \cong \overline{\mathrm{Hom}}_{kN_1}(U, V)$.

2 Let M be an A-module and N is a submodule with N the direct sum of its submodules V_1 and V_2, each isomorphic with the module V. If the extension of $N/V_1 \cong V$ by M/N is the same as the extension of $N/V_2 \cong V$ by M/N, in the sense of Proposition 7, then M has a quotient isomorphic to V.

22 Brauer graph

This section is devoted to the construction of the indecomposable projective B-modules. The detailed results of the preceding section are the essential tools including statements established during the course of proofs. In order to be able to use all the facts easily, we shall begin by displaying them in a useful table (Table 1), the first three columns related to L giving its dimension, the condition of Proposition 21.8 and the structure of L. The last four columns relate to the Green correspondent N of M, giving the structure of N, its socle, its radical quotient and a comparison of its composition length with that of V_{is} (so when L and M are projective there is no Green correspondent). The entries are clear from all the discussions of the preceding section, with the only exceptions being the first entries of the fifth and seventh columns. However, from the fourth column we have that $\mathrm{soc}(N) \cong V_{j+s+l(T_j)-q-1}$. But $j + l(T_j) - 1 \equiv \pi(j) = i$ (modulo e), as $\mathrm{soc}(T_j) \cong V_{\pi(j)}$, and we have always had $q \equiv 1$ (modulo e) so $\mathrm{soc}(N) \cong V_{i+s-1}$ as claimed. Finally, $l(N) = s + l(T_j) - q$, in this case, and $l(T_j) < q$, so $l(N) < s$.

We now set $\rho = \pi^{-1}$ and define a permutation σ by $\sigma(i) = \pi(i) + 1$, the composition of an e-cycle with π. We shall first construct a graph (which will turn out to be the tree we are after) associated with B. We start with a collection of e edges, one for each simple B-module, labelled accordingly. Each edge is to have two vertices (one at each end of the edge), one to be said to be of type ρ and one of type σ. For each orbit of ρ, we identify all the vertices of type ρ on the edges labelled by the simple modules S_i where i lies in the given orbit of ρ and we do the corresponding identification for σ; we get a well-defined graph. The edges around each vertex can be circularly ordered according to the orbit of ρ or σ that they correspond with and we shall assume this ordering is given. We shall devote the rest of the section to showing how the structure of this graph is closely related to the structure of the indecomposable projective B-modules. In the concluding sections we

Table 1

dim L	Condition	L	N	$\mathrm{soc}(N)$	$N/\mathrm{rad}(N)$	$l(N)$
$s+l(T_j)>q$	$i=\pi(j)$	$V_{j,s+l(T_j)-q} \oplus V_{i,q}$	$V_{j,s+l(T_j)}$	$\mathrm{soc}(V_{is}) \cong V_{i+s-1}$	$T_j/\mathrm{rad}(T_j) \cong V_j$	$l(N)<l(V_{is})=s$
$s+l(T_j)=q$	$i+s\equiv j$	V_{iq}	—	—	—	—
$s+l(T_j)<q$	$i+s\equiv j$	$V_{i,s+l(T_j)}$	$V_{i,s+l(T_j)}$	$\mathrm{soc}(T_j) \cong V_{\pi(j)}$	$V_{is}/\mathrm{rad}(V_{is}) \cong V_i$	$l(N)>l(V_{is})=s$

shall prove that the graph is a tree of the required sort. Notice now that the definition of the graph immediately implies that if we traverse a path in this graph then we shall pass vertices of type ρ and type σ in an alternative fashion.

We fix P_i as the indecomposable projective B-module corresponding with the simple B-module S_i and we can state the main result of the section.

Theorem 1 *The quotient* $\operatorname{rad}(P_i)/\operatorname{soc}(P_i)$ *is the direct sum of* (*possibly zero*) *uniserial modules* U_i^ρ *and* U_i^σ *with the following properties:*
  (1) *The factors of the radical series of* U_i^ρ *are*

$$S_{\rho(i)}, \ldots, S_{\rho^a(i)}$$

where $\rho^{a+1}(i) = i$;
  (2) *The factors of the radical series of* U_i^σ *are*

$$S_{\sigma(i)}, \ldots, S_{\sigma^{b+1}(i)}$$

where $\sigma^{b+1}(i) = i$;
  (3) U_i^ρ *and* U_i^σ *have no common composition factor.*

Notice that the third assertion is immediate: if it were otherwise then P_i would have a quotient module with a socle with a composition factor of multiplicity greater than one. Since any quotient of P_i is indecomposable, as $P_i/\operatorname{rad}(P_i)$ is simple, this violates Proposition 21.6. The non-negative integers a and b depend on i but we have suppressed this dependence to simplify our notation as we shall be concentrating on one module P_i. That the integers $a + 1$ and $b + 1$ are multiples (possibly greater than one) of the sizes of the orbits of $\langle \rho \rangle$ and $\langle \sigma \rangle$ on i is part of the statement of the theorem.

We shall proceed to prove the theorem by first stating and proving two results which construct large uniserial quotient modules of the projective module P_i.

Lemma 2 *There is a non-negative integer* b, *depending on* i, *such that* $\sigma^{b+1}(i) = i$ *and such that there is a uniserial module* W_i^σ *with radical series quotients*

$$S_i, S_{\sigma(i)}, \ldots, S_{\sigma^b(i)}$$

and Green correspondent $V_{i,q-1}$.

Lemma 3 *There is a non-negative integer* a, *depending on* i, *such that* $\rho^{a+1}(i) = i$ *and such that there is a uniserial module* W_i^ρ *with radical series* *quotients*

$$S_i, S_{\rho(i)}, \ldots, S_{\rho^a(i)}$$

and Green correspondent $V_{\pi(i)}$.

Let us begin by examining the situation for the permutation σ. If $l(T_i) + l(T_{\sigma(i)}) < q$ then there is an indecomposable module which is an extension of $S_{\sigma(i)}$ by S_i (see Table 1!) and has Green correspondent of length $l(T_i) + l(T_{\sigma(i)})$ with radical quotient isomorphic to V_i and socle isomorphic to $V_{\pi(\sigma(i))}$. If we have $l(T_i) + l(T_{\sigma(i)}) + l(T_{\sigma^2(i)}) < q$ then there is an extension of $S_{\sigma^2(i)}$ by the previously constructed module and this extension is indecomposable with Green correspondent of length $l(T_i) + l(T_{\sigma(i)}) + l(T_{\sigma^2(i)})$. Proceeding in this way, let b be the largest non-negative integer such that

$$l(T_i) + l(T_{\sigma(i)}) + \cdots + l(T_{\sigma^b(i)}) < b$$

and let U_i^σ be the indecomposable module obtained by the successive extensions.

Turning our attention to ρ, suppose that $l(T_i) + l(T_{\rho(i)}) > q$ so there is an indecomposable module which is an extension of $S_{\rho(i)}$ by S_i and has Green correspondent of length $l(T_i) + l(T_{\rho(i)}) - q$ with radical quotient isomorphic to $V_{\rho(i)}$ and socle isomorphic to $V_{\pi(i)}$. Now if $(l(T_i) + l(T_{\rho(i)}) - q) + l(T_{\rho^2(i)}) > q$ we can respect this process and construct an indecomposable module which is an extension of $T_{\rho^2(i)}$ by the previously constructed module and has Green correspondent of length $l(T_i) + l(T_{\rho(i)}) + l(T_{\rho^2(i)}) - 2q$. We can continue this process and, since each $\rho(T_j) < q$, there is a largest non-negative integer a such that

$$l(T_i) + l(T_{\rho(i)}) + \cdots + l(T_{\rho^a(i)}) - aq > q$$

and a corresponding indecomposable module which we shall take for U_i^σ.

In order to establish the stated properties of U_i^ρ and U_i^σ we begin with an easy result.

Lemma 4 *If M is a module with exactly three composition factors S_i, S_j, S_k, $0 \leq i, j, k \leq e - 1$ and $M/\mathrm{rad}(M)$ is simple then M is uniserial provided either $l(T_i) + l(T_j) + l(T_k) < q$ or $l(T_i) + l(T_j) + l(T_k) - 2q > q$.*

Proof Suppose that $M/\mathrm{rad}(M) \cong S_i$ so $\mathrm{rad}(M) \cong S_j \oplus S_k$ if the result is false. If the first inequality holds then the quotient of M by the submodule isomorphic with S_k is a non-split extension of S_j by S_i so $j = \sigma(i)$ as $l(T_j) + l(T_i) < q$. We similarly have $k = \sigma(i)$. The uniqueness of a non-split extension of $S_{\sigma(i)}$ by S_i now implies that $\mathrm{rad}(M) \cong S_{\sigma(i)}$, a contradiction. If the second inequality holds then a similar argument applies and the result is proved.

We now claim that U_i^q is uniserial. Our construction of U_i^q gives us a series of submodules

$$U_i^q = U_i^0 \supset U_i^1 \supset \cdots \supset U_i^a \supset U_i^{a+1} = 0$$

where $U_i^t/U_i^{t+1} \cong S_{\sigma^t(i)}$. The module U_i^0/U_i^2 is uniserial since it is a non-split of one simple module by another. We shall prove by induction that each quotient U_i^0/U_i^t is uniserial; since the idea can be seen in one case, we shall assume that U_i^0/U_i^a is uniserial but U_i^0 is not and derive a contradiction. In particular, the extension of U_i^a by U_i^{a-1}/U_i^a must split so $U_i^{a-1} \cong S_{\sigma^{a-1}(i)} \oplus S_{\sigma^a(i)}$. Applying the previous lemma to U_i^{a-2} we get that U_i^{a-2} has a quotient which is the direct sum of two simple modules. Continuing up the series given and applying the lemma repeatedly we get that $U_i^q/\mathrm{rad}(U_i^q)$ is the direct sum of two simple modules. This quickly implies that U_i^q has a summand isomorphic with $S_{\sigma^a(i)}$, contradicting the indecomposability of U_i^q, so U_i^q is as claimed. In an entirely similar manner, we also have that U_i^ρ is uniserial. It remains to establish the other assertions of Lemmas 2 and 3.

If V is the Green correspondent of U_i^σ then

$$\overline{\mathrm{Hom}}_{kG}(U_i^\sigma, S_{\sigma^{b+1}(i)}) \cong \overline{\mathrm{Hom}}_{kN_1}(V, T_{\sigma^{b+1}(i)}) \cong \overline{\mathrm{Hom}}_{kN_1}(\Omega V, \Omega T_{\sigma^{b+1}(i)}).$$

But $l(V) + l(T_{\sigma^{b+1}(i)}) \geqslant q$, by our choice of b, so $l(\Omega V) + l(\Omega T_{\sigma^{b+1}(i)}) \leqslant q$ and thus, by Lemma 21.3,

$$\overline{\mathrm{Hom}}_{kN_1}(\Omega V, \Omega T_{\sigma^{b+1}(i)}) \cong \mathrm{Hom}_{kN_1}(\Omega V, \Omega T_{\sigma^{b+1}(i)}).$$

Now

$$\Omega V/\mathrm{rad}(\Omega V) \cong V_{\pi(\sigma^b(i))+1} = V_{\sigma^{b+1}(i)}$$

while

$$\mathrm{soc}(\Omega T_{\sigma^{b+1}(i)}) \cong T_{\sigma^{b+1}(i)}/\mathrm{rad}(T_{\sigma^{b+1}(i)}) \cong V_{\sigma^{b+1}(i)}$$

so certainly $\mathrm{Hom}_{kN_1}(\Omega V, \Omega T_{\sigma^{b+1}(i)}) \neq 0$. Thus, $\mathrm{Hom}_{kG}(U_i^\sigma, S_{\sigma^{b+1}(i)}) \neq 0$ so, as $U_i^\sigma/\mathrm{rad}(U_i^\sigma) \cong V_i$ we have $\sigma^{b+1}(i) = i$. Now $\mathrm{soc}(V) \cong V_{\pi(\sigma^b(i))}$ and $i = \sigma^{b+1}(i) = \pi(\sigma^b(i)) + 1$ so $\mathrm{soc}(V) \cong V_{i-1}$. Hence, $l(V) = q - 1 - me, m \geqslant 0$. If $m > 0$ and T_i is short then

$$l(V) + l(T_{\sigma^{b+1}(i)}) = l(V) + l(T_i) \leqslant (q - 1 - e) + e < q$$

which contradicts our choice of b. If $m > 0$ and T_i is long then

$$l(V) \geqslant l(T_i) \geqslant q - e > q - 1 - me = l(V),$$

again a contradiction. Hence, $m = 0$, $l(V) = q - 1$ and therefore $V \cong V_{i,q-1}$ and the second lemma is fully proved.

Now let V be the Green correspondent of U_i^ρ. We have

$$\overline{\mathrm{Hom}}_{kG}(U_i^\rho, S_{\rho^{a+1}(i)}) \cong \overline{\mathrm{Hom}}_{kN_1}(V, T_{\rho^{a+1}(i)}) \cong \mathrm{Hom}_{kN_1}(V, T_{\rho^{a+1}(i)})$$

since $l(V) + l(T_{\rho^{a+1}(i)}) \leqslant q$ by our choice of a. But

$$\mathrm{soc}(T_{\rho^{a+1}(i)}) V_{\pi(\rho^{a+1}(i))} = V_{\rho^a(i)}$$

as $\pi = \rho^{-1}$, while $V/\mathrm{rad}(V) \cong V_{\rho^a(i)}$, so certainly $\mathrm{Hom}_{kN_1}(V, T_{\rho^{a+1}(i)}) \neq 0$ and we deduce that $\rho^{a+1}(i) = i$. Now

$$\mathrm{soc}(V) \cong \mathrm{soc}(V_i) \cong V_{\pi(i)} = V_{\rho^{-1}(i)}$$

$$V/\mathrm{rad}(V) \cong T_{\rho^a(i)}/\mathrm{rad}(T_{\rho^a(i)}) \cong V_{\rho^a(i)}$$

so $\mathrm{soc}(V) \cong V/\mathrm{rad}(V)$ and thus $l(V) = me + 1$, $m \geqslant 0$. If $m > 0$ and T_i is long then

$$l(V) + l(T_{\rho^{a+1}(i)}) = l(V) + l(T_i) \geqslant e + 1 - q - e > q,$$

which contradicts our choice of a. If $m > 0$ and T_i is short then, since $l(V) < l(T_i)$, we have

$$e + 1 \leqslant l(V) < l(T_i) \leqslant e$$

which is also a contradiction. Hence, $m = 0$, $V \cong V_{\pi(i)}$ and the third lemma is proved.

Let us now turn to the proof of the theorem.

Proof Since U_i^ρ and U_i^σ have isomorphic radical quotients we can construct a module U which has a submodule M which is the direct sum of modules M_ρ and M_σ and such that $U/M \cong S_i$, $M_\rho \cong \mathrm{rad}(U_i^\rho)$, $M_\sigma = \mathrm{rad}(U_i^\sigma)$, $U/M_\sigma \cong U_i^\rho$, $U/M_\rho \cong U_i^\sigma$. We claim that U has the structure claimed for $P_i/\mathrm{soc}(P_i)$ in the theorem. It clearly suffices to show that $M = \mathrm{rad}(U)$. Since this is clear if $M_\rho = 0$ or $M_\sigma = 0$ we assume that both these modules are non-zero. Let $\bar{U} = U/\mathrm{rad}(M)$ and use bars to denote images in $\bar{U}$ so that $\bar{M} = \bar{M}_\rho + \bar{M}_\sigma$ is a direct sum, $\bar{M}_\rho \cong S_{\rho(i)}$, $\bar{M}_\sigma \cong S_{\sigma(i)}$, $\bar{U}/\bar{M}_\sigma$ is a non-split extension and so is $\bar{U}/\bar{M}_\rho$. The condition on length of Green correspondents in extensions of simple modules guarantees that $S_{\rho(i)} \not\cong S_{\sigma(i)}$ and so $\bar{M}_\rho$, $\bar{M}_\sigma$ and 0 are the only proper submodules of M. Since each of them has its quotient in $\bar{U}$ not semisimple we are left with $\mathrm{rad}(\bar{U}) = \bar{M}$ as desired.

Hence, it remains only to prove that there is an extension of S_i by U which is projective. We can regard U as obtained by iterated extensions, starting with U_i^σ and successively using $S_{\rho(i)}$, $S_{\rho^2(i)}$, ... and so on. Since $U/\mathrm{rad}(U)$ is simple each of the successive extensions is an indecomposable module. Moreover, since the Green correspondent of U_i^σ has length $q - 1$, which is at least as large as the length of the correspondent T_i of S_i, each of these successive extensions must be of 'ρ-type'. Hence, we deduce that the Green correspondent V of U has $V/\mathrm{rad}(V) \cong V_{\rho^a(i)} = V_{\pi(i)}$. Moreover, the Green correspondents of U_i^ρ and U_i^σ have lengths 1 and $q - 1$ so $l(V) \equiv 1 + (q - 1) - l(T_i)$ (modulo q) so $l(V) = q - l(T_i)$. But $V/\mathrm{rad}(V) \cong V_{\pi(i)}$ so V and

$\Omega^{-1}(T_i)$ have the same length and isomorphic radical quotients. Thus $V \cong$ $\Omega^{-1}(T_i)$ and, in particular, $\mathrm{soc}(V) \cong V_{i-1}$. But now we have all the conditions needed to guarantee that there is an extension of S_i by U which is projective.

We now wish to make a useful technical extension of the idea of a Brauer tree and a Brauer tree algebra. We simply drop the conditions that the graph be a tree, that is, have no cycles, and that there be exactly one vertex with a multiplicity. We allow cycles and attach multiplicities to each vertex and have the idea of a *Brauer graph*; using this notion to describe algebras exactly as we did before we have the concept of a *Brauer graph algebra*. We can now state the final result of the section.

Theorem 5 *The block B is a Brauer graph algebra.*

We have described a graph, at the beginning of the section; in view of Theorem 1, all that remains is to assign multiplicities to the vertices. Let us consider the edge labelled by the simple module S_i and its vertex which is of type ρ. In order to satisfy the requirements stated in the theorem, we need to attach a multiplicity of $(a+1)/t$, where a is as in Theorem 1 and depends on i and t is the size of the orbit of i under the action of $\langle \rho \rangle$. Hence, we need to see this is independent of which edge we are considering attached to the one vertex of type ρ. Once we see this (and since the analogous situation will clearly hold for σ) the theorem will be proved in all parts.

However, $\mathrm{rad}(U_i^\rho)$ is a homomorphic image of $P_{\rho(i)}$ and so it is a homomorphic image of $U_{\rho(i)}^\sigma$, since $U_{\rho(i)}^\sigma$ and $U_{\rho(i)}^\rho$ share exactly one composition factor. Thus, $l(U_i^\rho) \leqslant l(U_{\rho(i)}^\rho)$ and, continuing in this fashion, using the circularity of the ordering, we get that all the inequalities are equalities and all is proved.

Exercises

1 Let $U_1, \ldots, U_r$ be the indecomposable B-modules, one of each isomorphism type. Let $W_1, \ldots, W_r$ be their Green correspondents so $\overline{\mathrm{Hom}}_{kG}(U_i, U_j) \cong \overline{\mathrm{Hom}}_{kN_1}(W_i, W_j)$ and ΩU_i has correspondent ΩW_i. Set $X_i = \Omega W_i$ and show the 'correspondence' which attached X_i to U_i has the same properties.

2 With the 'new' correspondence show that the 'roles' of ρ and σ are interchanged.

3 Any non-projective B-module obtained from S_i by iterated extensions of 'type σ' is a homomorphic image of U_i^σ.

23 Trees

The main result of this section brings us almost to the goal of the entire chapter.

Theorem 1 *The block B is a Brauer tree algebra.*

Once this is established, it will remain only to determine the multiplicity, which will be done in the next and last section. Since B is a block, it is immediate that the corresponding Brauer graph that we have constructed is connected. Hence, we need to show that it is a tree, that is, has no cycles, and that all the multiplicities are one, with at most a single exception.

The technical result that we require to carry this out is as follows.

Lemma 2 *Let M be an A-module and N a submodule satisfying the following conditions:*
(1) *$M/\mathrm{rad}\,M$ is multiplicity-free;*
(2) *$N/\mathrm{rad}\,N$ is simple;*
(3) *$N \not\subseteq \mathrm{rad}\,M$;*
(4) *There is a semisimple submodule S of $\mathrm{rad}\,M$ such that $M/\mathrm{rad}\,M$ and $\mathrm{rad}\,M/S$ have no common composition factor.*
It follows that if $M = M_1 + M_2$ is a direct decomposition then there is $i, i = 1$ or 2, such that $M_i \supset N \cap S$ and M_i has a quotient isomorphic with $N/\mathrm{rad}(N)$.

This result will be used to prove that certain modules are indecomposable by virtue of their having many submodules to which the lemma can be applied.

Proof Let $N_i, i = 1, 2$, be the projection of N into the summand M_i. We may assume that $N_1 \neq 0$ and $N_2 \neq 0$ since otherwise N is contained in M_1 or M_2 and we are done. But N_1 and N_2 are homomorphic images of N so (2) now implies that

$$N_1/\mathrm{rad}\,N_1 \cong N_2/\mathrm{rad}\,N_2 \cong N/\mathrm{rad}\,N.$$

By choice of notation and (3), we have $N_1 \not\subseteq \mathrm{rad}\,M_1$. Thus, $M_1/\mathrm{rad}\,M_1$ has a composition factor isomorphic with $N/\mathrm{rad}\,N$ and so $M_2/\mathrm{rad}\,M_2$ does not, by (1). Hence, $N_2 \subset \mathrm{rad}\,M_2 \subset \mathrm{rad}\,M$ so $N_1 \subset S$, by (4). Since S is semisimple we deduce that N_2 is simple. Hence, since $N \cap S$ is a proper submodule of N, we have

$$N \cap S \subset \mathrm{rad}\,N \subset \mathrm{rad}\,N_1 + \mathrm{rad}\,N_2 = \mathrm{rad}\,N_1 \subset M_1,$$

and the lemma is proved.

Let us now turn to the proof of the theorem. First, suppose that the graph is not a tree so it contains a cycle. Choose a cycle of minimal length. Since there are two kinds of vertices, the cycle has even length. Moreover, this length is at least four; otherwise it is two and so consists of two edges with endpoints being the same two vertices. In this case, the recipe for the structure of indecomposable projective modules readily implies that each of the indecomposable projective modules corresponding to the edges in the cycle has a quotient with a socle which is not multiplicity-free, which is a contradiction.

Hence, we can assume that we have a cycle of length $2r$ with the edges labelled by the distinct simple modules $R_1, \ldots, R_{2r}$ in that order. Moreover, the minimality of the cycle guarantees that the vertex attached to R_i, R_{i+1} is not attached to the edge labelled by any other R_j. In particular, the recipe for the structure of projective modules implies that the indecomposable projective module corresponding with R_i, $1 \leqslant i \leqslant 2r$, has a quotient module N_i satisfying the following conditions: $N_i/\text{rad } N_i c R_i$; $\text{soc}(N_i) \cong R_{i-1} \oplus R_{i+1}$; $\text{rad} N_i/\text{soc } N_i$ has no composition factor isomorphic with any R_j. (Here, as usual, subscripts are modulo $2r$ so $R_{2r+1} = R_1$.) Now consider the following direct sum:

$$N_1 \oplus N_3 \oplus \cdots N_{2r-1}.$$

We have $\text{soc}(N_1) \cong R_{2r} \oplus R_2$, $\text{soc}(N_3) \cong R_2 \oplus R_4$ and so on until $\text{soc}(N_{2r-1}) \cong R_{2r-2} \oplus R_{2r}$. We now wish to form a quotient of this direct sum in which we 'identify' the submodule of N_1 isomorphic to R_2 with the submodule of N_3 isomorphic to R_2, the submodule of N_3 isomorphic to R_4 with the submodule of N_5 isomorphic to R_4, and so on. In fact, let V_{2i}, $1 \leqslant i \leqslant r-1$, be a submodule of the direct sum isomorphic to R_{2i} which projects onto the simple submodules of N_{2i-1} and N_{2i+1} isomorphic to R_{2i} and which has zero projection on every other N_j is the sum. We now let M be the quotient of the direct sum by the sum of all the V_{2i}; since each N_j is isomorphic with its image in M, we may assume $N_j \subseteq M$ and deduce that M has the following properties, where we set $S = \text{soc}(N_1) \oplus \cdots \oplus \text{soc}(N_{2r-1})$:

(1) $M = N_1 + \cdots + N_{2r-1}$;
(2) $M/\text{rad } M \cong R_1 \oplus \cdots \oplus R_{2r-1}$;
(3) $\text{rad } M \supseteq S \cong R_{2r} \oplus R_2 \oplus \cdots \oplus R_{2r-2} \oplus R_{2r}$;
(4) $\text{rad } M/S$ has no composition factor isomorphic with any R_j;
(5) $N_{2i-1} \cap N_{2i+1} \cap S \cong R_{2i}$.

We can picture M schematically then as follows:

$$
\begin{array}{ccccccc}
R_1 & & R_3 & & & R_{2r-1} & \\
\diagup \;\; \diagdown & \diagup \;\; \diagdown & \diagup \cdots \diagdown & & \diagup \;\; \diagdown \\
R_{2r} & R_2 & R_4 & R_{2r-2} & & R_{2r}
\end{array}
$$

We now claim that M is indecomposable. Indeed, the lemma implies that if M is the direct sum of two modules then one of them contains $N_1 \cap S$ and has $R_1 \cong N_1/\text{rad } N_1$ as a quotient. Similarly, the lemma also forces one summand to contain $N_3 \cap S$ and have R_3 as a quotient. But $N_1 \cap N_3 \cap S \neq 0$ so the same summand has R_1 and R_3 as quotients. Continuing in this fashion, this one summand has the same radical quotient as M so it is M and M is as asserted. But $\text{soc}(M)$ contains S and is therefore not multiplicity-free, so we have a contradiction and the graph is a tree as claimed above.

Next, suppose that there are two vertices with a multiplicity greater than one and choose a path of minimal length in the tree connecting two such vertices. Suppose that this path is labelled as follows:

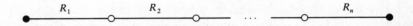

As before, each vertex pictured has no edge attached to it labelled with any R_j except for the displayed ones. We let N_1 be a quotient of the projective indecomposable corresponding to R_1 which has socle isomorphic to $R_1 \oplus R_2$ and rad $N_1/\text{soc } N_1$ having no composition factor isomorphic to R_j; we can do this as the left-hand vertex has multiplicity greater than one. We now choose N_i, $1 < i < n$, with $\text{soc}(N_i) \cong R_{i-1} \oplus R_{i+1}$, in a similar fashion and finally N_n with $\text{soc}(N_n) \cong R_{n-1} \oplus R_n$. Notice that $n > 1$ or else N_1 would be indecomposable and $\text{soc}(N_1)$ would not be multiplicity-free.

We must now consider two cases, depending on the parity of n. If n is odd form

$$N_1 \oplus N_3 \oplus \cdots \oplus N_n \oplus N_{n-1} \oplus \cdots \oplus N_2$$

and pass to a quotient M, by identification, so that $M = N_1 + \cdots + N_n + N_{n-1} + \cdots + N_2$ and so that, schematically, we have the following picture of M:

$$
\begin{array}{ccccccccccc}
R_1 & & R_3 & & & & R_n & & R_{n-1} & & R_2 \\
\diagup\;\diagdown & & \diagup\;\diagdown & & \diagup\cdots\diagdown & & \diagup\;\diagdown & & \diagup & \diagdown\cdots & \diagup\;\diagdown \\
R_1 & & R_2 & & R_4 & & R_{n-1} & & R_n & & R_1
\end{array}
$$

The argument used above applies again and we deduce that M is indecomposable and $\text{soc}(M)$ is not multiplicity-free. If n is even, we consider

$$N_1 \oplus N_3 \oplus \cdots \oplus N_{n-1} \oplus N_n \oplus N_{n-2} \oplus \cdots \oplus N_2$$

and we get a module M, pictured as follows:

$$
\begin{array}{ccccccccccc}
R_1 & & R_3 & & R_{n-1} & & R_n & & & & R_2 \\
\diagup\;\diagdown & & \diagup\;\diagdown\cdots\diagup & & \diagdown & & \diagup\;\diagdown & & \diagup\cdots\diagdown & & \diagup\;\diagdown \\
R_1 & & R_2 & & & & R_n & & R_{n-1} & & R_3 & & R_1
\end{array}
$$

This again yields a contradiction in the same way so we deduce that there is at most one vertex which has a multiplicity exceeding one. Hence, the theorem is proved.

Exercises

1   Show that the lemma holds equally well for decomposition into more than two summands.
2   Devise an argument that deals with cycles of odd length and avoids referring to the two types of vertices.

24 Multiplicity

We have reached our last task, the determination of the exceptional multiplicity of B. We begin by introducing some ideas in the case of the block b_1. We let $G_0(b_1)$ be the free abelian group (written additively) with a basis consisting of the symbols $[V_1], \ldots, [V_e]$, one for each simple b_1-module. If V is any b_1-module and the multiplicity of V_i as a composition factor of V is a_i then we set

$$[V] = a_1[V_1] + \cdots + a_e[V_e].$$

In particular if

$$0 \to U \to V \to W \to 0$$

is an exact sequence then

$$[V] = [U] + [W].$$

Let $K_0(b_1)$ be the subgroup of $G_0(b_1)$ spanned by all $[Q]$, where Q is a projective b_1-module, and set $\overline{G_0(b_1)} = G_0(b_1)/K_0(b_1)$.

Lemma 1 *The group $\overline{G_0(b_1)}$ is finite of order p^n.*

Proof Let $Q_i = V_{iq}$, $1 \leqslant i \leqslant e$, be the indecomposable projective module corresponding with V_i. Therefore, $K_0(b_1)$ is spanned by $[Q_1], \ldots, [Q_e]$. Let

$$[Q_i] = c_{i1}[V_1] + \cdots + c_{ie}[V_e]$$

so, if $C = (c_{ij})$ (the Cartan matrix of b_1), then, by the theory of elementary divisors, the group $\overline{G_0(b_1)}$ is finite if, and only if, $\det(C) \neq 0$ and when finite has order $|\det(C)|$. It remains therefore only to calculate this determinant.

However, b_1 is a Brauer tree algebra of a special sort and we know the structure of each Q_i. Hence, we readily have, where $\mu = (p^n - 1)/e$,

$$
C = \begin{pmatrix} \mu+1 & \mu & \cdots & \mu \\ \mu & \mu+1 & \cdots & \mu \\ \vdots & & & \vdots \\ \mu & \mu & \cdots & \mu+1 \end{pmatrix}.
$$

Thus, $C = \mu J + I$, where J is the e by e matrix with all entries equal to one and I is the e by e identity matrix. Hence, $C = (-\mu)(-J - (1/\mu)I)$ so $\det(C)$ is the product of $(-\mu)^e$ and the value of the characteristic polynomial of $-J$ at $1/\mu$. But $-J$ has eigenvalues $0, 0, \ldots, 0, -e$ (as J has rank one and the column vectors of all ones is an eigenvector for eigenvalue $-e$ for $-J$) so

$$
\det(C) = (-\mu)^e (-1/\mu)^{e-1}(-e - 1/\mu) = e\mu + 1 = p^n,
$$

as asserted.

The abelian groups $G_0(B)$, $K_0(B)$ and $\overline{G_0(B)}$ can now be defined in exactly the same fashion. The critical relation is the following:

Lemma 2 *The groups $\overline{G_0(B)}$ and $\overline{G_0(b_1)}$ are isomorphic.*

If U is a B-module then $b_1 U$ is a b_1-module, the summand of U_{N_1} lying in the block b_1. Since $b_1 U$ is also the product of the identity element of b_1 with U, it follows that if

$$
0 \to U \to V \to W \to 0
$$

is an exact sequence, then so is

$$
0 \to b_1 U \to b_1 V \to b_1 W \to 0
$$

Hence, there is a homomorphism of $G_0(B)$ to $G_0(b_1)$ which sends $[U]$ to $[b_1 U]$. Moreover, if U is projective then so is U_{N_1} and also $b_1 U$ so we have a homomorphism of $\overline{G_0(B)}$ to $\overline{G_0(b_1)}$. Similarly, using induction, we have a homomorphism of $\overline{G_0(b_1)}$ to $\overline{G_0(B)}$. Finally, if U is a non-projective indecomposable B-module and V is the corresponding b_1-module then $b_1 U$ is the direct sum of V and a projective module while BV^G is the direct sum of U and a projective module, so the compositions of the two homomorphisms constructed, in either order, is the identity and the lemma is proved.

Therefore, now let C be the Cartan matrix of the block B. Let m be the exceptional multiplicity of B.

Lemma 3 *The determinant of C is $em + 1$.*

Note that this is all we need as Lemmas 1 and 2 will then imply that

$em + 1 = p^n$ and $m = (p^n - 1)/e$. Let us begin by describing C in detail. The simple B-modules are $S_1, \ldots, S_e$ ($S_e \cong S_0$). The i, j entry of the Cartan matrix of B is the multiplicity of S_j as a composition factor of the indecomposable projective module corresponding with S_i. The description of the projective modules in terms of the tree immediately implies that this i, j entry is the number of vertices, counting multiplicity, in common between the edge labelled with S_i and the edge labelled with S_j.

First, suppose that the tree for B is a star. If the exceptional vertex is at the center then the determinant of C is $em + 1$, as the argument in the proof of Lemma 1 shows after replacing μ with m. If the exceptional vertex is not at the center, so the tree has the form

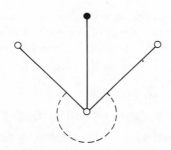

then

$$C = \begin{pmatrix} m+1 & 1 & \cdots & 1 \\ 1 & 2 & & 1 \\ \vdots & & & \vdots \\ 1 & 1 & \cdots & 2 \end{pmatrix}.$$

Let us prove a result stronger than the one stated in Lemma 3, that is, that the Cartan matrix, defined in terms of common vertices, of any tree with an exceptional vertex with multiplicity m and e edges has determinant $em + 1$. We have already done this in the case where the exceptional vertex is in the center. Now the 1, 1 minor of C is the Cartan matrix of a tree with $e - 1$ edges and exceptional multiplicity one so this minor has determinant e so that $\det(C) = (m + 1)e + r$, where r is the sum of the remaining terms in the first row cofactor expansion of C. However, if $m = 1$ we already have that the determinant of C is $e + 1$ so

$$e + 1 = 2e + r$$

and $r = -e + 1$ yielding $(m + 1)e + r = em + 1$. In order to deal with the case of general trees, and not just stars, we need an easy result about determinants which generalizes the fact that the determinant of a matrix in block form is the product of the determinants of the blocks. Suppose that M is a square matrix and

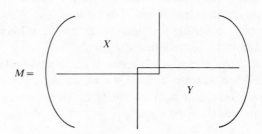

where X and Y are square and intersect in a single entry which is z. Moreover, assume that the entries of M outside X and Y are zero. Let X_0 be the minor of X obtained by deleting the last row and column of X and let Y_0 be the minor obtained from Y by deleting the first row and column of Y.

Lemma 4 *We have the equality*

$$\mathrm{Det}(M) = \mathrm{Det}(X)\,\mathrm{Det}(Y_0) + \mathrm{Det}(X_0)\,\mathrm{Det}(Y) - \mathrm{Det}(X_0)z\,\mathrm{Det}(Y_0).$$

This is immediate from the Laplace expansion for determinants or is easily verified directly. We can now complete the proof of Lemma 3. We proceed by induction on the number of edges and we can assume that the tree is not a star. In this case, choose an edge which is not an end so the tree is the union of two trees which intersect in the edge. We have the following picture, where the displayed edge is the selected one and the vertices may or may not be exceptional:

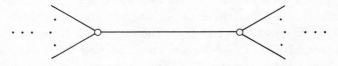

Let L be the tree on the left, R the one on the right, so they intersect in the edge E pictured. Let L_0 and R_0 be the tree obtained by deleting the edge E from L and R, respectively. The previous lemma applies immediately to the Cartan matrix $C(T)$ of the tree T in question and we have

$$\mathrm{Det}\,C(T) = \mathrm{Det}\,C(L) \cdot \mathrm{Det}\,C(R_0) + \mathrm{Det}\,C(L_0) \cdot \mathrm{Det}\,C(R)$$
$$- \mathrm{Det}\,C(L_0) \cdot z \cdot \mathrm{Det}\,C(R_0)$$

where z is the sum of the multiplicities of the two vertices of E (that is, 2 or

$m + 1$ according as neither is exceptional or one of them is). By symmetry, we can assume the exceptional vertex is in L and there are two cases, as it is a vertex of E or is not. In each case, a simple calculation, using induction on L, L_0, R, R_0, gives the desired result and all, at long last, is proved.

A guide to further reading

This volume has emphasized depth by proceeding directly towards the most important results but this has been at the expense of breadth and breadth is an essential part of the subject. We have relied exclusively on module-theoretic methods but ring-theoretic methods (e.g., idempotents, ideals, trace maps) and character-theoretic methods (e.g., characters, Brauer characters, decomposition numbers) are vital and complete results cannot be obtained without them. The interested reader should now (or while reading the book) broaden his knowledge and this guide is designed to be of assistance.

Our first chapter treats the Wedderburn theorems, and their consequences for group representations, in a special case, that of algebras over algebraically closed fields. The results extend to algebras over arbitrary fields (an easy generalization), to rings with minimum condition and, to a good extent, to arbitrary rings. The treatise of Curtis and Reiner [3] treats all of this but only the case of algebras over arbitrary fields has been used in studying representations of finite groups.

The next three chapters can be seen as a basic course in module-theoretic methods in group representations. The seventh chapter of the Huppert–Blackburn treatise [7] has a similar purpose, intersects and complements our coverage and has many good exercises. On the other hand, Goldschmidt's book [5] gives an up-to-date treatment of how characters can be used directly; this amounts to an approach that is the most distant from ours. Landrock's book [8] shows how the ring-theoretic methods can be used and can be thought of as intermediate between character and module methods. Of course, Feit's comprehensive treatise [4] covers all approaches but is best used by a reader with some previous background. Reading some of these sources will give the reader a better idea of the full picture: representations in characteristic zero, characteristic p and over valuation rings are studied and related by use of three different methods while proving theorems connecting a group and its p-local subgroups.

The final section of Chapter 4 is an introduction to methods which synthesize notions from the structure theory of groups and representation theory. Building on these ideas, with ring-theoretic methods, Broué and Puig [1, 2] have developed a theory of nilpotent blocks which is the only rival of the cyclic theory of Chapter 5 for completeness and depth. It is to be hoped that further research will lead to results which encompass both these deep theories as special cases.

The fifth and final chapter is devoted to the cyclic theory. The only complete treatment in book form is in Feit's treatise [4] but Goldschmidt's book [5] and Green's [6] examine special cases from different points of view. The process of reduction modulo p which is essential to applications to character values is not touched here but the books of Curtis–Reiner [3], Green [6] and Serre [9] give good introductory treatments. We have kept the algebraic prerequisites for this chapter limited but a knowledge of more algebra would lead to some shortening of proofs. For example, the last part of section 19 is devoted to giving a Morita equivalence between two algebras (that is, showing their module categories are equivalent). The general theory of such equivalences is treated in Curtis–Reiner [3]. In section 20 some knowledge of homological algebra would help. In particular, use of the functor Ext and dimension shifting would greatly shorten the proof of Lemma 6.

In any case, we hope we have given the reader some idea of how to explore the subject a little more using the base we have supplied.

Bibliography

The books in the list below all contain extensive lists of references to the published research literature.

1   M. Broué and L. Puig, Characters and local structure in G-algebras, *J. Algebra* **63** (1980), 306–17.

2   M. Broué and L. Puig, A Frobenius theorem for blocks, *Inv. Math.* **56** (1980), 117–28.

3   C. W. Curtis and I. Reiner, *Methods of Representation Theory*, vol. I, John Wiley and Sons, 1981.

4   W. Feit, *The Representation Theory of Finite Groups*, North-Holland, 1982.

5   D. M. Goldschmidt, *Lectures on Character Theory*, Publish or Perish, 1980.

6   J. A. Green, *Vorlesungen über modulare Darstellungstheorie endlicher Gruppen*, Vorlesungen aus dem Mathematischen Institut Giessen, Heft 2, 1974.

7   B. Huppert and N. Blackburn, *Finite Groups* II, Grundlehren der mathematischen Wissenschaften 242, 1982.

8   P. Landrock, *Finite Group Algebras and their Modules*, London Mathematical Society Lecture Note Series 84, Cambridge University Press, 1983.

9   J.-P. Serre, *Linear Representations of Finite Groups*, Graduate Texts in Mathematics 42, Springer-Verlag, 1977.

Index